ANNALES ACADEMIÆ SCIENTIARUM FENNICÆ

MATHEMATICA

DISSERTATIONES

144

THE GROUP OF ORBIT PRESERVING G-HOMEOMORPHISMS OF AN EQUIVARIANT SIMPLEX FOR G A LIE GROUP

ROMAN GOEBEL

University of Helsinki, Department of Mathematics and Statistics

To be presented, with the permission of the Faculty of Science of the University of Helsinki, for public criticism in Auditorium XII, the Main Building of the University, on October 15th, 2005, at 10 a.m.

HELSINKI 2005
SUOMALAINEN TIEDEAKATEMIA

ISSN 1239-6303
ISBN 951-41-0976-7

Received 15 September 2005

2000 Mathematics Subject Classification:
Primary 57S05; Secondary 57S20, 54E35, 54B15.

YLIOPISTOPAINO
HELSINKI 2005

Acknowledgements

I would like to express my gratitude to my instructor, Professor Sören Illman, for introducing me to the subject, giving me the freedom to invent and providing me the working conditions for this thesis.

Also, I feel that it is appropriate to acknowledge the collective efforts of the mathematical community and those supporting it. The rather altruistic efforts of past mathematicians constitute a basis for this thesis and even for my understanding of mathematics. For financial support and educational efforts I am indepted to the Graduate School of Mathematical Analysis and Logic and the Magnus Ehrnrooth Foundation.

These past years I have enjoyed the company of Erik Elfving, Tuomas Korppi, Anssi Lahtinen and Elena Ravaioli and many others who have attended the Transformation Groups Seminar with me. I wish to thank you all for the opportunity to present and refine my ideas you have given me.

I would like to thank Docent Jouni Luukkainen and Professor Alexandr Mischenko for conducting a detailed preliminary examination of this thesis in a convenient schedule and my father Robert for help with language.

I would like to thank my friends and colleagues and the helpful staff of the Department.

My parents and grandmother have supported me in many ways during my studies. The most grateful I am to my wife Maija and son Ralf who have offered me a perfect balance between real life and its "strech, bend, but please don't tear" topological idealisation.

I wish to dedicate this thesis to Elsi Carlstedt, my mathematics teacher during 1988–1991.

Helsinki, 15 September 2005

Roman Goebel

Contents

Introduction

Let G be a Lie group and let $K_0 \supset \cdots \supset K_n$ be compact subgroups. The n-dimensional *equivariant simplex* $\Delta_n(G; K_0, \ldots, K_n)$, introduced in [Ill72b], [Ill72a] and [Ill83], is a special sort of G-space whose orbit space is a simplex Δ_n and whose orbit structure is closely connected with the structure of the simplex.

The main application of equivariant simplexes is in *equivariant triangulations* of G-spaces. Such theorems concern the question of the *existence* of a triangulation of the orbit space M/G of a G-space M such that each part of M that lies over a simplex of M/G is G-homeomorphic to an equivariant simplex $\Delta_n(G; K_0, \ldots, K_n)$ for some choice of $K_0 \supset \cdots \supset K_n$. The existence of an equivariant triangulation of a smooth G-manifold M was shown in [Ill78] for a finite discrete group G and in [Ill83] for a compact Lie group G, see also [Ill72b] and [Ill72a]. For a convenient account of the present status, including the concept of *uniqueness*, of equivariant triangulations of smooth proper G-manifolds for *any* Lie group G, the reader is referred to [Ill02].

An equivariant triangulation of a G-space M gives M a *G-CW complex structure* in a natural way, see Proposition II.2.3 in [Ill72a] and also Proposition 6.1 in [Ill83]. Indeed, for G a compact Lie group and M a smooth G-manifold, a G-CW complex structure is, for practical purposes, unique if it is to induce a subanalytic triangulation on the orbit space M/G, see [MS87], Theorem 4.2, for the precise meaning. While the main motivation for this work lies elsewhere, the results of this thesis are incidentally linked to the cited *uniqueness* result of G-CW complex structures, see Section 9.3 and our remarks below.

In this thesis we study deformations of the group of orbit preserving G-homeomorphisms of a fixed equivariant simplex $\Delta_n(G; K_0, \ldots, K_n)$, denoted henceforth by X. Although the focus in the literature has been on the left G-action $\Phi\colon G \times X \to X$, the equivariant simplex X also exhibits a natural right N_0^n-action $\Psi\colon X \times N_0^n \to X$ which commutes with the left G-action and preserves G-orbits. Here for $k = 0$ and below for any $0 \leq k \leq n$ we denote by N_k^n the intersection $N(K_k) \cap \cdots \cap N(K_n)$ of normalisers.

The group of orbit preserving (i.e., over Δ_n) G-homeomorphisms of X, denoted by $\mathbf{Homeo}^G_{\Delta_n}(X)$, is given the compact-open topology. It is well-known that it is a topological group when G is a Lie group. Since the right N^n_0-action Ψ commutes with the left G-action and preserves G-orbits, the adjoint of Ψ is an antihomomorphism $\psi\colon N^n_0 \to \mathbf{Homeo}^G_{\Delta_n}(X)$. Furthermore, ψ is a proper map and the kernel of ψ is K_n. Hence the image $\psi(N^n_0)$ is a closed subgroup of $\mathbf{Homeo}^G_{\Delta_n}(X)$, isomorphic to N^n_0/K_n.

The main point demonstrated in this thesis is that *the group $\psi(N^n_0)$ is a strong equivariant deformation retract of* $\mathbf{Homeo}^G_{\Delta_n}(X)$ when the subgroup $\psi(N^n_0)$ acts on $\mathbf{Homeo}^G_{\Delta_n}(X)$ by multiplication from the right, see Theorem 8.5. It follows that $\mathbf{Homeo}^G_{\Delta_n}(X)$ has the homotopy type of the Lie group N^n_0/K_n when G is a Lie group. As a simple consequence of equivariance we also mention that

1. $\psi(N^n_0)$ has a global cross-section in $\mathbf{Homeo}^G_{\Delta_n}(X)$; and
2. $\mathrm{id}\,\psi(N^n_0)$ is a strong deformation retract of $\mathbf{Homeo}^G_{\Delta_n}(X)/\psi(N^n_0)$.

Even though the literature on equivariant simplexes centers on the case that G is a Lie group, equivariant simplexes may be of interest more generally. To show that $\psi(N^n_0)$ is a strong equivariant deformation retract of $\mathbf{Homeo}^G_{\Delta_n}(X)$, it suffices to assume that G is a metrisable group and that N^n_{k-1} is a strong equivariant neighbourhood deformation retract of N^n_k, see Theorem 8.4. An example, in which G is not even locally compact, is given in Example 9.9.

The author was led to work on the questions of this thesis when Sören Illman requested him to show that any orbit preserving G-homeomorphism of X is G-homotopic over Δ_n to an element in $\psi(N^n_0)$ when G is a Lie group. The idea of developing this result in the more general setting of function spaces was partly influenced by a result coined the *Straightening Lemma*, see [Bre72], V.4.1. The equivariance aspect simply emerged from the equations during the work. The same applies to the non-Lie setting.

The structure of the thesis is as follows. Part I explains what an equivariant simplex is. In Section 1 we recall some basic results about topological groups, proper group actions and proper maps, and this will allow us to fix our terminology. In Section 2 we define the equivariant simplexes and prove their basic properties. We conclude Part I by showing that X is G-invariantly metrisable whenever G is a metrisable group.

In Part II we work with the G-space $\mathbf{CS}(X)$ formed by the cross-sections of the equivariant simplex X and prove alternative versions of our main results. We now assume that G is a metrisable group. The equivariant simplex X contains a subset X_n such that $\mathbf{CS}(X_n)$ is a topological group with a neutral element η. The set $\mathbf{CS}(X_n)$ also exhibits a natural *left* action of the group N^n_0 which is compatible with the group law and whose ineffective kernel is K_n. Theorem 7.2 then states that under the same assumptions on the normalisers $N(K_0), \ldots, N(K_n)$ as referred to above, the orbit $N^n_0 \cdot \eta$ is a strong N^n_0-equivariant deformation retract of $\mathbf{CS}(X_n)$. By an easy application of well-known results concerning Lie groups we conclude Theorem 7.3. The connection between Theorems 7.2 and 8.4 and between Theorems 7.3 and 8.5 is given by anti-isomorphic homeomorphisms

$$\mathbf{CS}(X_n) \underset{\eta^*}{\overset{\epsilon}{\rightleftarrows}} \mathbf{Homeo}^G_{\Delta_n}(X),$$

but the treatment of this correspondence will be left until Section 8.

Part II begins with Section 3, where well-known statements about the compact-open topology are presented to facilitate efficient treatment of function spaces. In Section 4 we define an increasing sequence X_k, $0 \le k \le n$, of N_0^n-invariant subsets of X over Δ_n and see that this gives rise to an increasing sequence of N_0^n-invariant subsets $\mathbf{CS}(X_k)$ of $\mathbf{CS}(X)$. In Theorem 4.3 we show that $\mathbf{CS}(X_n)$ is a topological group. Section 4 is concluded with Proposition 4.4, which shows how certain types of parametrised deformations of the simplex Δ_n, the domain of the cross-sections, induce deformations of the spaces of cross-sections. The assumptions of Proposition 4.4 also illustrate the main obstacle encountered in this approach to constructing deformations of spaces of cross-sections: the deformation does not allow the desired "dimension drop".

In Section 5 we discuss on a general level the conditions and implications of when a subgroup M is an M-equivariant (neighbourhood) deformation retract of a topological group N. Theorem 5.1 provides a satisfactory characterisation of this situation.

Sections 6 and 7 form the core of this thesis. There is an intricate interplay of topology and algebra which makes it possible to define a set W_k such that $X_{k-1} \subset W_k \subset X_k$ and such that there exists an N_0^n-map $H_k \colon I \times W_k \to X_k$ over Δ_n which deforms W_k into X_{k-1}, see Theorem 6.4. The set W_k has the feature that there exists an N_0^n-map $\hat{R}_k \colon I \times \mathbf{CS}(X_k) \to \mathbf{CS}(X_k)$ which deforms $\mathbf{CS}(X_k)$ into $\mathbf{CS}(W_k)$, see Proposition 6.8. The dimension drop is then handled by deforming the range W_k into X_{k-1}. This leads inductively to an N_0^n-deformation of $\mathbf{CS}(X_n)$ into $\mathbf{CS}(X_0)$, see Corollary 6.10. An easy application of Proposition 4.4 yields an N_0^n-deformation of $\mathbf{CS}(X_0)$ onto the orbit $N_0^n \cdot \eta$, see Theorem 7.1.

In Part III we focus on the group $\mathbf{Homeo}^G_{\Delta_n}(X)$. In Section 8 the results obtained for $\mathbf{CS}(X_n)$ in Part II are translated to $\mathbf{Homeo}^G_{\Delta_n}(X)$ using the correspondence mentioned above, see Theorem 8.3. As a byproduct we obtain the comforting result that $\mathbf{Homeo}^G_{\Delta_n}(X)$ is a topological group. The group $\psi(N_0^n)$ is seen to correspond to $N_0^n \cdot \eta$.

The thesis is concluded with Section 9 where we discuss some implications and applications of the main results. In Section 9.1 we briefly discuss some implications of the main results to the structure of $\mathbf{Homeo}^G_{\Delta_n}(X)$ as already mentioned above. In Section 9.2 we show how the Straightening Lemma can be obtained as a special case of our main theorem.

In Sections 9.3 and 9.4 we apply our results to solve two less general, but still very difficult problems. In Section 9.3 we show that if α and β are two cross-sections of the equivariant simplex X having a certain special isotropy structure, then there exists a $g \in N_0^n$ such that α and $g\beta$ are connected by an isotropy structure preserving homotopy of cross-sections. For G a compact Lie group, this result is recorded in Lemma 4.4 in [MS87] and it is used to establish the uniqueness of a G-CW structure on a C^k G-manifold M, see Theorem 4.2 in [MS87]. However, it seems that the proof of Lemma 4.4 is not correctly formulated, as we will point out. In Section 9.4 we make an analogous statement and show that every G-homeomorphism $h\colon X \to X$ over Δ_n is homotopic through G-homeomorphisms over Δ_n to some element in the group $\psi(N_0^n)$ of right translations. As mentioned above, this was the original motivation for the work.

Finally, in Section 9.5 we give for each $n \geq 1$ an example of a non-locally compact group G and a non-trivial sequence of compact subgroups $K_0 \supset \cdots \supset K_n$ for which the assumptions of Theorems 7.2 and 8.4 are satisfied. Since every Lie group is locally compact, this shows that the applicability of the main results is not limited to Lie groups.

Part I. Equivariant simplexes: basic properties and G-invariant metrisation

To avoid cumbersome notation, throughout this thesis maps that are obtained by restricting the domain or range of a given map are denoted using the same letter as the original map, see Proposition 1.9, for example.

1. Background on proper group actions

Throughout Section 1 we assume that G is a topological group. The neutral element of G is denoted by e.

The purpose of this section is to recall basic information about topological groups and group actions. Familiar constructions like homogeneous spaces appear as examples for a reason: they are 0-dimensional equivariant simplexes.

1.1. Topological groups and group actions

By a *topological group* we mean a group G such that the set G has a topology which makes the multiplication $(g, g') \mapsto gg'$ and the inverse element map $g \mapsto g^{-1}$ continuous.

By an *action* of G on a topological space X we mean a function

$$\Phi\colon G \times X \to X$$

such that

1. $\Phi(g', \Phi(g, x)) = \Phi(g'g, x)$ for all g', g and x; and
2. $\Phi(e, x) = x$ for all x.

To be precise, the above definition concerns a *left* action. By a *right* action we mean a function $\Psi\colon X \times G \to X$ such that $\Phi\colon G \times X \to X$, $(g,x) \mapsto \Psi(x, g^{-1})$, is a left action.

A space X together with a *continuous* action is called a *G-space*. It will be convenient to denote gx for $\Phi(g,x)$.

Example 1.1. Let $K \subset G$ be a subgroup. The rule $(k,g) \mapsto gk^{-1}$ makes G a left K-space and the rule $(g,k) \mapsto gk$ makes G a right K-space. □

Example 1.2. Let Y be a topological space. The projection $\mathrm{pr}\colon G \times Y \to Y$ is a left action, the *trivial action*. If X is a G-space, then the rule $(g,(x,y)) \mapsto (gx,y)$ makes $X \times Y$ a G-space. □

Two points x and y in a G-space X are said to be on same *orbit* if there exists a $g \in G$ such that $y = gx$. Every G-space decomposes into disjoint orbits Gx and the orbit space X/G is given the *quotient topology* induced by the *orbit map* $\pi\colon X \to X/G$, $x \mapsto Gx$; a set U is open in X/G if and only if $\pi^{-1}(U)$ is open in X. It is easy to prove that the orbit map is an open map, see [Bre72], p. 37.

Let X and Y be G-spaces. A (continuous) map $f\colon X \to Y$ is said to be *G-equivariant* or a *G-map* if $f(gx) = gf(x)$ for all g and x. A G-map $f\colon X \to Y$ induces a unique continuous map $\tilde{f}\colon X/G \to Y/G$ satisfying $\tilde{f}\pi = \pi f$.

If Y is merely a topological space, then a map $f\colon X \to Y$ is said to be *G-invariant* if $f(gx) = f(x)$ for all g and x. This only means that f is G-equivariant if one considers Y as a trivial G-space. Note also that a G-invariant map $f\colon X \to Y$ induces a unique continuous map $\tilde{f}\colon X/G \to Y$ satisfying $\tilde{f}\pi = f$.

Proposition 1.3. *Let X be a G-space and let Y be another topological space on which G acts and let $p\colon X \to Y$ be a surjective G-map such that* $\mathrm{id} \times p\colon G \times X \to G \times Y$ *is a quotient map. Then also Y is a G-space.*

Proof. Let $\Phi\colon G \times X \to X$ and $\Psi\colon G \times Y \to Y$ denote the actions. Since p is a G-map, it satisfies $p \circ \Phi = \Psi \circ (\mathrm{id} \times p)$. By assumption $\mathrm{id} \times p$ is a quotient map, and since $p \circ \Phi$ is continuous, it follows that Ψ is continuous. □

Example 1.4. Let $K \subset G$ be a subgroup. In both cases that appear in Example 1.1, the orbit space is the *homogeneous space* G/K, the elements of which are the left cosets of K in G, and the orbit map is the canonical map $\pi\colon G \to G/K$.

Note that

$$\begin{array}{rccc} \Phi\colon & G \times G/K & \to & G/K \\ & (g, g'K) & \mapsto & gg'K \end{array}$$

gives a left G-action on G/K. The space G/K also exhibits a right action of the normaliser of K. The *normaliser* of K in G is the subgroup

$$N(K) = \{\ g \in G \mid gKg^{-1} = K\ \},$$

and

$$\begin{aligned} \Psi\colon \quad G/K \times N(K) &\rightarrow G/K \\ (gK, n) &\mapsto gnK \end{aligned}$$

gives a well-defined right $N(K)$-action. For if $gK = g'K$, then $g'^{-1}g \in K$ and hence $(g'n)^{-1}(gn) = n^{-1}g'^{-1}gn \in K$, so that $gnK = g'n'K$. These actions *commute* in the sense that $(gx)n = g(xn)$ holds for all $g \in G$, $x \in G/K$ and $n \in N(K)$.

Recall that G may also be considered as a left G-space and a right $N(K)$-space, see Example 1.1. The canonical map $\pi\colon G \to G/K$ is then a left G-map and a right $N(K)$-map. Since π is open and surjective, so are $\mathrm{id} \times \pi$ and $\pi \times \mathrm{id}$. Hence both Φ and Ψ are continuous by Proposition 1.3. □

Each point x of a G-space X can be associated with a subgroup

$$G_x = \{ g \in G \mid gx = x \}$$

of G, the *isotropy subgroup* of x. If x and y are on the same orbit, then there exists a $g \in G$ such that $y = gx$ and one can show that $G_y = gG_xg^{-1}$. This shows that the isotropy subgroups of points on one orbit form the conjugacy class of G_x in G.

By a *G-homeomorphism* $f\colon X \to Y$ we mean a G-map which is also a homeomorphism. It follows that the inverse map $f^{-1}\colon Y \to X$ is then also a G-map. A G-homeomorphism $f\colon X \to Y$ is *isovariant*, which means that $G_{f(x)} = G_x$ for all $x \in X$.

Let X be a G-space. A subset $A \subset X$ is said to be *G-invariant* if $gx \in A$ for all g and all $x \in A$. We obtain a natural *induced action* $\Phi\colon G \times A \to A$. It is easy to see that the *interior* $\mathrm{int}\, A$ and the *closure* $\overline{A}$ of a G-invariant set A are G-invariant.

Let X be a G-space, let H be a topological group and let $\varphi\colon H \to G$ be a continuous homomorphism. Then the rule $(h, x) \mapsto \varphi(h)x$ shows how H can act on X *through* φ. A commonly encountered example is that H is a subgroup of G and φ is the inclusion.

A group action is said to be *effective* if $gx = x$ for all x implies that $g = e$. Subgroups that act by left or right multiplication, for example, are always effective. The *ineffective kernel* of an action is

$$N = \{ g \in G \mid gx = x \text{ for all } x \in X \},$$

which is always a normal subgroup of G. There is a natural induced action of G/N on X which is continuous and effective, see [Bre72], Proposition I.1.1.

1.2. Embeddings and quotient maps

Let $f\colon X \to Y$ be a map. If f is injective and each open $U \subset X$ can be expressed as $U = f^{-1}(V)$ for some open $V \subset Y$, we say that f is an *embedding*. If f is surjective and $f^{-1}(V) \subset X$ is open only for open $V \subset Y$, we say that f is a *quotient map*. Embeddings and quotient maps have the following characteristic feature, which was used in Section 1.1.

Proposition 1.5. *Let X, Y and Z be any spaces, and let $f\colon X \to Y$ and $g\colon Y \to Z$ be functions.*

1. *If g is an embedding and gf is continuous, then f is continuous.*
2. *If f is a quotient map and gf is continuous, then g is continuous.* □

Example 1.6. Let X, Y and Y' be G-spaces and let $f\colon X \to Y$ and $f'\colon X \to Y'$ be G-maps and quotient maps. If the fibres of f and f' coincide (i.e., $f^{-1}(f(x)) = f'^{-1}(f'(x))$ for all $x \in X$), then there exists a unique bijective function $h\colon Y \to Y'$ satisfying $hf = f'$ and h is a G-homeomorphism. □

The following simple result provides, in some situations, a very efficient way to show that a map is an embedding or a quotient map.

Proposition 1.7. *Let X, Y and Z be any spaces, and let $f\colon X \to Y$ and $g\colon Y \to Z$ be maps.*

1. *If gf is an embedding, then so is f.*
2. *If gf is a quotient map, then so is g.* □

Every open (or closed) injective map is an embedding and every open (or closed) surjective map is a quotient map.

Proposition 1.8. *Let $f\colon X \to Y$ be a quotient map. Then $f\colon f^{-1}(B) \to B$ is a quotient map for every open $B \subset Y$.*

Proof. Let $U \subset B$ be such that $f^{-1}(U)$ is open in $f^{-1}(B)$. Since B is open in Y, $f^{-1}(B)$ is open in X, and hence $f^{-1}(U)$ is open in X. Note finally that since $U \subset B$, the set $f^{-1}(U)$ is the same as for the original f. Since the original f was a quotient map, U is open in Y and hence also in B. □

1.3. Proper maps

The (cartesian) product of two open maps is an open map, but closed maps have the interesting feature that the product of two closed maps is not necessarily closed. However, if the maps have compact point-inverses, the situation is totally different.

By a *proper* map we mean a map $f\colon X \to Y$ which satisfies either of the following two conditions which are shown equivalent in [Bou66], Theorem I.10.1.

1. For every topological space Z the map $\mathrm{id} \times f\colon Z \times X \to Z \times Y$ is a closed map.
2. The map f is closed and for every point $y \in Y$ the inverse image $f^{-1}(y)$ is compact.

The composition of two proper maps is a proper map. It is a simple consequence of 1 that for any space Z the map $\mathrm{id} \times f$ is again a proper map. Hence if $f\colon X \to Y$ and $h\colon Z \to W$ are two proper maps, then the product

$$f \times h = (f \times \mathrm{id}) \circ (\mathrm{id} \times h)$$

is also proper.

The following proposition will be used frequently.

Proposition 1.9. *Let $f\colon X \to Y$ be a proper map. Then*

1. *$f\colon A \to f(A)$ is proper for every closed $A \subset X$; and*
2. *$f\colon f^{-1}(B) \to B$ is proper for every $B \subset Y$.*

Proof. For 1, see [Bou66], I.10, Propositions 2, 5a) and 5c). For 2, see [Bou66], I.10, Proposition 3a). □

The following is a useful result concerning closed functions.

Lemma 1.10. *Let $f\colon X \to Y$ be a continuous map.*

1. *If f is closed, then for each $y \in Y$ and each neighbourhood U of the fibre $f^{-1}(y)$ there exists a neighbourhood W of y such that $f^{-1}(W) \subset U$.*
2. *If f is a quotient map and if for each $y \in Y$ and each neighbourhood U of the fibre $f^{-1}(y)$ there exists a neighbourhood W of y such that $f^{-1}(W) \subset U$, then f is closed.*

Proof. For the proof of the first statement, let f be closed, let $y \in Y$ and let U be an open neighbourhood of the fibre $f^{-1}(y)$. Then $X \setminus U$ is closed, and therefore $W = Y \setminus f(X \setminus U)$ is an open neighbourhood of y. It is easy to see that $f^{-1}(W) \subset U$.

For the proof of the second statement, let $F \subset X$ be closed and assume that f is a quotient map satisfying the given condition. Let $y \in Y \setminus f(F)$. Now $U = X \setminus F$ is a neighbourhood of the fibre $f^{-1}(y)$, so by assumption, y has a neighbourhood W such that $f^{-1}(W) \subset U$. It is easy to see that $W \cap f(F) = \emptyset$, and hence $f(F)$ is closed. □

Examples of non-trivial proper maps are easily constructed using the following result.

Proposition 1.11. *Let K be a compact group and let X be a K-space. Then the action map $\Phi\colon K \times X \to X$ and the orbit map $\pi\colon X \to X/K$ are both proper maps and, in particular, closed maps.*

Proof. For the first assertion, see [Bou66], Proposition III.4.1.

To see that the orbit map π is a proper map, we show that it is closed and has compact fibres. That the fibres are compact is evident since they are of the form Kx. Let $F \subset X$ be closed. To see that $\pi(F)$ is closed it suffices to notice that the set $\pi^{-1}(\pi(F)) = KF$ is closed since Φ is a closed map. □

Let us end this section with an important basic result that J. L. Kelley credits to A. D. Wallace, see [Kel55], Theorem 5.12. It states that each neighbourhood of a compact cartesian set in a product space contains a cartesian neighbourhood.

Lemma 1.12 (Wallace). *Let X and Y be spaces, $A \subset X$ and $B \subset Y$ compact sets and W a neighbourhood of $A \times B$ in $X \times Y$. Then A has a neighbourhood U and B a neighbourhood V such that $U \times V$ is contained in W.* □

1.4. Twisted products and cones

In this section we recall two familiar constructions, the twisted product and the cone.

Let G be a topological group and let $K \subset G$ be a subgroup. If X is a left K-space, then K acts on $G \times X$ by $(k,(g,x)) \mapsto (gk^{-1}, kx)$. The underlying space of the *twisted product* $G \times_K X$ is defined as the K-orbit space $(G \times X)/K$. The orbits, that is, the values of the natural projection

$$q\colon G \times X \to G \times_K X$$

are denoted by $q(g,x) = [g,x]$. Note that $[g,x] = [g',x']$ if and only if $g'^{-1}g \in K$ and $x' = g'^{-1}gx$. Note also that q is always open and that q is proper if K is compact, by Proposition 1.11.

The group G acts on $G \times X$ in a natural way by $g(g',x) = (gg',x)$ and this action induces an action on X that has the form

$$\begin{array}{rccc} \Phi\colon & G \times (G \times_K X) & \to & G \times_K X \\ & (g,[g',x]) & \mapsto & [gg',x] \end{array}.$$

Since $\mathrm{id}_G \times q$ is an open map and hence a quotient map, Φ is continuous by Proposition 1.3.

Let G be a topological group and let X be a *compact* G-space. The underlying space of the *cone* $C(X)$ is defined as the quotient space $I \times X/\sim$ where

$$(t,x) \sim (t',x') \quad \text{if } t = t' \text{ and } x = x'; \text{ or if } t = 0 = t'.$$

The equivalence classes, that is, the values of the natural projection

$$q\colon I \times X \to (I \times X)/\sim,$$

are denoted by $q(t,x) = [t,x]$. Since the saturation of a closed set $F \subset I \times X$ is F or $F \cup 0 \times X$, q is a closed map. Since X is compact, all fibres are compact, and hence q is a proper map.

The group G acts on $I \times X$ in a natural way by $g(t,x) = (t,gx)$ and this action induces an action on $C(X)$ that has the form

$$\begin{array}{rccc} \Phi\colon & G \times C(X) & \to & C(X) \\ & (g,[t,x]) & \mapsto & [t,gx] \end{array}.$$

Since $\mathrm{id}_G \times q$ is a proper map and hence a quotient map, Φ is continuous by Proposition 1.3.

1.5. Proper actions

The notion of a proper action is not central to the main results of this thesis. However, it provides a straightforward method for proving that the transformations form a closed subgroup of the function space, see Sections 4.4 and 8.3. Moreover, equivariant simplexes and proper actions appear frequently together in the literature.

The literature knows various non-equivalent ways to define a proper action. The definition adopted here is used in [Bou66], Chapter III.4. We say that a continuous action $\Phi\colon G \times X \to X$ is *proper* and X is a *proper* G-space if the map

$$\begin{aligned} \Phi'\colon \quad G \times X &\to X \times X \\ (g, x) &\mapsto (gx, x) \end{aligned}$$

is a proper map.

Proposition 1.13. *Let X be a proper G-space and let $x \in X$. Then the map*

$$\begin{aligned} \lambda\colon \quad G &\to X \\ g &\mapsto gx \end{aligned}$$

is a proper G-map, which induces a G-homeomorphism between the homogeneous space G/G_x and the orbit Gx, which is closed in X.

Proof. It is plain that λ is a G-map. For the remaining assertions, see [Bou66], Proposition III.4.4. □

Example 1.14. Let G be a topological group. Multiplication $\mu\colon G \times G \to G$ makes G a proper left G-space. Namely, $\mu'\colon G \times G \to G \times G$, $(g, g') \mapsto (gg', g')$ is a homeomorphism whose inverse is $(g, g') \mapsto (gg'^{-1}, g')$.

Similar remarks apply to the right action as well. □

Example 1.15. Let A be any space. Then G acts continuously on $G \times A$ by $\Phi : (g, (g', a)) \mapsto (gg', a)$. If A is Hausdorff, this action is also proper as we are about to show.

Let Δ denote the diagonal map

$$\begin{aligned} \Delta\colon \quad A &\to A \times A \\ a &\mapsto (a, a) \end{aligned}.$$

Since $\mathrm{pr} \circ \Delta = \mathrm{id}$ and since $A \times A$ is a Hausdorff space, it follows from Proposition I.10.5d) in [Bou66] that Δ is a proper map.

We now consider the map

$$\begin{aligned} \Phi'\colon \quad G \times (G \times A) &\to (G \times A) \times (G \times A) \\ (g, (g', a)) &\mapsto ((gg', a), (g', a)) \end{aligned}.$$

By selecting a suitable permutation σ of $(G \times A) \times (G \times A)$ one may write $\sigma \circ \Phi' = \mu' \times \Delta$. This shows that $\sigma \circ \Phi'$ is a proper map, and consequently so is Φ'.

Similar remarks apply to the right action as well. □

Example 1.16. Let G be a topological group and let X be a proper G-space with an action Φ. If H is another topological group and if $\varphi\colon H \to G$ is a homomorphism and a proper map, then the action

$$\begin{aligned} \Psi\colon \quad H \times X &\to X \\ (h, x) &\mapsto \varphi(h)x \end{aligned}$$

makes X a proper H-space. For note that $\Psi' = \Phi' \circ (\varphi \times \mathrm{id})$. In particular, for any proper G-space X and any closed subgroup H, the restricted action of H is a proper action. □

Proposition 1.17. *Let X be a G-space and let Y be another topological space on which G acts and let $p\colon X \to Y$ be a surjective proper G-map. Then*

1. *G acts continuously on Y; and*
2. *Y is a proper G-space if X is a proper G-space.*

Proof. Let $\Phi\colon G \times X \to X$ and $\Psi\colon G \times Y \to Y$ denote the actions.

1. Since p is surjective and proper, so is $\mathrm{id} \times p$, and hence $\mathrm{id} \times p$ is a quotient map. Hence we may use Proposition 1.3.
2. For the proof that Ψ is a proper action, see [Bou66], Proposition III.4.5(i). □

The next result concerns closed subgroups in a general topological group G.

Proposition 1.18. *Let G be a topological group and let $K \subset G$ be a closed subgroup. Then $N(K)$ is closed in G and G/K is Hausdorff.*

Proof. For the first assertion, let $g \notin N(K)$. Then there exists a $k \in K$ such that $gkg^{-1} \notin K$ or $g^{-1}kg \notin K$. Suppose, for example, that the former holds. Since $g \mapsto gkg^{-1}$ is continuous and since $G \setminus K$ is open, g has a neighbourhood U such that $g'kg'^{-1} \in G \setminus K$ for all $g' \in U$. Clearly $U \subset G \setminus N(K)$. A similar argument can be used for $g \mapsto g^{-1}kg$ as well. Therefore, $N(K)$ is closed in G.

For the second assertion, see [Bou66], Proposition III.2.13. □

Example 1.19. Let G be a topological group and let K be a compact subgroup. The homogeneous space G/K has a left G-action and a right $N(K)$-action, see Example 1.4. Moreover, if one considers G as a left G-space and a right $N(K)$-space, then the canonical map $\pi\colon G \to G/K$ is a left G-map and a right $N(K)$-map.

According to Example 1.14, G is a proper left G-space. Since π is a surjective proper map by Proposition 1.11, it follows from Proposition 1.17.2 that G/K is a proper left G-space.

If K is *closed* in G, then also $N(K)$ is closed in G by Proposition 1.18. It follows from Examples 1.14 and 1.16 that G is a proper right $N(K)$-space. Hence also G/K is a proper right $N(K)$-space if K is closed in G. □

The following basic results are found in [Bou66], Proposition III.4.5(ii) and Example III.4.1.2).

Proposition 1.20. *Let X be a Hausdorff G-space and let Y be a proper G-space.*

1. *If there exists a G-map $f\colon X \to Y$, then G acts properly on X.*
2. *If $B \subset Y$ is a G-invariant subspace, then G acts properly on B.* □

2. Equivariant simplexes

Throughout Section 2 we assume that G is a topological group and $K_0 \supset \cdots \supset K_n$ are compact subgroups.

In this section we define the equivariant simplex $\Delta_n(G; K_0, \ldots, K_n)$ in a very general setting. The main result of Section 2.2 is Theorem 2.1 concerning basic properties of the equivariant simplex. For another way of achieving similar results under the assumption that G is locally compact and Hausdorff, see [Ill00], for example. In Section 2.3 we show that up to a G-homeomorphism, the equivariant simplex X can be constructed using cones and twisted products.

In Section 2.4 we show that if G is metrisable, then the topology of X is induced by a G-invariant metric.

2.1. The standard simplex

Using the vector space structure of $\mathbb{R}^{n+1}$ we define the *standard n-simplex* to be the set

$$\Delta_n = \{ \ (a_0, \ldots, a_n) \in \mathbb{R}^{n+1} \mid a_0 + \cdots + a_n = 1, a_0, \ldots, a_n \geq 0 \ \}.$$

We also adopt the convention that $\Delta_{-1} = \emptyset$. The *m-vertex* of Δ_n is the point $\mathbf{e}_m$ given by the $(m+1)^{\text{th}}$ standard basis vector. We denote $(a_0, \ldots, a_n)$ by $\boldsymbol{a}$.

For $0 \leq m \leq n$ we consider Δ_{m-1} as a subset of Δ_m through the *front face imbedding*

$$\begin{array}{rccc} i_m\colon & \Delta_{m-1} & \to & \Delta_m \\ & \boldsymbol{a} & \mapsto & (a_0, \ldots, a_{m-1}, 0). \end{array}$$

This induces a filtration $\Delta_{-1} \subset \cdots \subset \Delta_n$ of the standard n-simplex. The filtration in turn induces a natural partition of Δ_n by the sets $\Delta_m \setminus \Delta_{m-1}$ $(0 \leq m \leq n)$. Naturally, for each $\boldsymbol{a} \in \Delta_n$ there is a unique m between 0 and n such that $\boldsymbol{a} \in \Delta_m \setminus \Delta_{m-1}$. We use this to define the *dimension function*

$$\begin{array}{rccc} \delta\colon & \Delta_n & \to & \{0, \ldots, n\} \\ & \boldsymbol{a} & \mapsto & \max\{ \ 0 \leq m \leq n \mid a_m > 0 \ \} \end{array}.$$

2.2. The equivariant simplex

The underlying space of the *n-dimensional equivariant simplex* $\Delta_n(G; K_0, \ldots, K_n)$ is defined as the quotient space $G \times \Delta_n / \sim$ where

$$(g, \boldsymbol{a}) \sim (g', \boldsymbol{a}') \quad \text{if } \boldsymbol{a} = \boldsymbol{a}' \text{ and } gK_{\delta(\boldsymbol{a})} = g'K_{\delta(\boldsymbol{a}')}.$$

The equivalence classes, that is, the values of the natural projection

$$p\colon G \times \Delta_n \to G \times \Delta_n / \sim,$$

are denoted by $p(g, \boldsymbol{a}) = \langle g, \boldsymbol{a} \rangle$. To shorten the notation, we denote the space $\Delta_n(G; K_0, \ldots, K_n)$ by X. The natural projection $p\colon G \times \Delta_n \to X$ is in fact a proper map, see Theorem 2.1.6 below.

The group G acts on $G \times \Delta_n$ in a natural way by $g(g', \boldsymbol{a}) = (gg', \boldsymbol{a})$, and this action induces an action on X that has the form

$$\begin{array}{cccc} \Phi\colon & G \times X & \to & X \\ & (g, \langle g', \boldsymbol{a}\rangle) & \mapsto & \langle gg', \boldsymbol{a}\rangle \end{array}.$$

This action is both continuous and proper, see Theorem 2.1.8 below, and the natural projection $p\colon G \times \Delta_n \to X$ is G-equivariant. Observe that points $\langle g, \boldsymbol{a}\rangle$ and $\langle g', \boldsymbol{a}'\rangle$ are on the same G-orbit if and only if $\boldsymbol{a} = \boldsymbol{a}'$. The isotropy subgroup G_x at $x = \langle g, \boldsymbol{a}\rangle$ is $gK_{\delta(\boldsymbol{a})}g^{-1}$.

The map $\mathrm{pr}\colon G \times \Delta_n \to \Delta_n$ factors through p inducing a well-defined map

$$\begin{array}{cccc} \varpi\colon & X & \to & \Delta_n \\ & \langle g, \boldsymbol{a}\rangle & \mapsto & \boldsymbol{a} \end{array},$$

which is G-invariant. Observe that

$$\begin{array}{cccc} \eta\colon & \Delta_n & \to & X \\ & \boldsymbol{a} & \mapsto & \langle e, \boldsymbol{a}\rangle \end{array}$$

is a map such that $\varpi\eta = \mathrm{id}$, and hence ϖ is a quotient map and η is an embedding by Proposition 1.7. Observe that the fibres of ϖ are exactly the G-orbits of X. Therefore, ϖ factors through the orbit space X/G inducing a homeomorphism $\tilde{\varpi}$. From hereon we consider Δ_n as the orbit space of X and $\varpi\colon X \to \Delta_n$ as the orbit map. Hence $\eta\colon \Delta_n \to X$ is a global cross-section and we call it the *neutral cross-section*. Using ϖ we may also extend the function δ to $\delta\colon X \to \{0, \dots, n\}$.

Let us adopt the following notation. If $A \subset \Delta_n$ and $X' \subset X$, then by $X'|A$ we mean the space $\varpi^{-1}(A) \cap X'$. For example, $X|(\Delta_n \setminus \Delta_{m-1})$ is an open G-invariant subspace of X such that $p^{-1}(X|(\Delta_n \setminus \Delta_{m-1})) = G \times (\Delta_n \setminus \Delta_{m-1})$. Therefore, the restricted G-map

$$p_m\colon G \times (\Delta_n \setminus \Delta_{m-1}) \to X|(\Delta_n \setminus \Delta_{m-1})$$

is a quotient map by Proposition 1.8.

Let $\pi_m\colon G \to G/K_m$ denote the canonical map. Observe next that the G-map

$$\pi_m\,\mathrm{pr}\colon G \times (\Delta_n \setminus \Delta_{m-1}) \to G/K_m$$

factors through p_m inducing a G-map

$$\begin{array}{cccc} \gamma_m\colon & X|(\Delta_n \setminus \Delta_{m-1}) & \to & G/K_m \\ & \langle g, \boldsymbol{a}\rangle & \mapsto & gK_m \end{array}.$$

Using ϖ and γ_m we can construct a G-equivariant map

$$\begin{array}{cccc} \omega_m\colon & X|(\Delta_n \setminus \Delta_{m-1}) & \to & G/K_m \times (\Delta_n \setminus \Delta_{m-1}) \\ & \langle g, \boldsymbol{a}\rangle & \mapsto & (gK_m, \boldsymbol{a}) \end{array},$$

which is given by (γ_m, ϖ). Also important is the G-map

$$\begin{array}{rccc} e_m\colon & G/K_m \times (\Delta_m \setminus \Delta_{m-1}) & \to & X|(\Delta_n \setminus \Delta_{m-1}) \\ & (gK_m, \boldsymbol{a}) & \mapsto & \langle g, \boldsymbol{a}\rangle \end{array}.$$

Notice that $\omega_m e_m = \mathrm{id}$, and hence e_m is an embedding by Proposition 1.7.1. Moreover, ω_m is a proper map, and hence a quotient map. Namely, it holds that $\omega_m \circ p_m = \pi_m \times \mathrm{id}$. Since π_m is proper by Proposition 1.11, so is $\pi_m \times \mathrm{id}$, and since p_m is surjective, ω_m is proper by [Bou66], Proposition I.10.5b).

Let N_0^n denote the group $N(K_0) \cap \cdots \cap N(K_n)$. It has a natural right action $((g, \boldsymbol{a}), n) \mapsto (gn, \boldsymbol{a})$ on $G \times \Delta_n$. Similarly, the equivariant simplex exhibits a right N_0^n-action

$$\begin{array}{rccc} \Psi\colon & X \times N_0^n & \to & X \\ & (\langle g, \boldsymbol{a}\rangle, n) & \mapsto & \langle gn, \boldsymbol{a}\rangle \end{array}.$$

Let us show that this is well-defined. Let $n \in N_0^n$ and let $\langle g, \boldsymbol{a}\rangle = \langle g', \boldsymbol{a}\rangle$, which means that $g'^{-1}g \in K_{\delta(\boldsymbol{a})}$. Since $n \in N_{\delta(\boldsymbol{a})}$, it follows that $n^{-1}g'^{-1}gn \in K_{\delta(\boldsymbol{a})}$, and hence $\langle gn, \boldsymbol{a}\rangle = \langle g'n, \boldsymbol{a}\rangle$. This action is continuous, see Theorem 2.1.9 below, and the natural projection $p\colon G \times \Delta_n \to X$ is right N_0^n-equivariant. Observe also that Ψ commutes with the left G-action and that Ψ preserves G-orbits.

Theorem 2.1. *Let X denote the equivariant simplex $\Delta_n(G; K_0, \ldots, K_n)$.*

1. *The space Δ_n and the map $\varpi\colon X \to \Delta_n$ can be considered as the orbit space and the orbit map of X.*
2. *The restricted map $p_m\colon G \times (\Delta_n \setminus \Delta_{m-1}) \to X|(\Delta_n \setminus \Delta_{m-1})$ is a quotient map.*
3. *The map $\omega_m\colon X|(\Delta_n \setminus \Delta_{m-1}) \to G/K_m \times (\Delta_n \setminus \Delta_{m-1})$ is proper.*
4. *The coset map $\gamma_m\colon X|(\Delta_n \setminus \Delta_{m-1}) \to G/K_m$ is a continuous G-map.*
5. *If U is a neighbourhood of x in X and $m = \delta(x)$, then there exists a neighbourhood W of $\omega_m(x)$ in $G/K_m \times (\Delta_n \setminus \Delta_{m-1})$ such that ${\omega_m}^{-1}(W) \subset U$.*
6. *The quotient map $p\colon G \times \Delta_n \to X$ is a proper map.*
7. *X is Hausdorff if $K_0, \ldots, K_n$ are closed in G.*
8. *X is a proper left G-space.*
9. *X is a right N_0^n-space. The action is proper if $K_0, \ldots, K_n$ are closed in G.*

Proof. Statements 1–4 have been discussed above.

5. Let U be a neighbourhood of x. Let $m = \delta(x)$. We may assume that $U \subset X|(\Delta_n \setminus \Delta_{m-1})$ and so we may make use of the proper map ω_m. Since ${\omega_m}^{-1}(\omega_m(x)) = \{x\}$, U is a neighbourhood of the fibre. Hence it follows from Lemma 1.10.1 that there exists a neighbourhood W of $\omega_m(x)$ such that ${\omega_m}^{-1}(W) \subset U$.
6. It is clear that p has compact fibres and hence it suffices to show that p is closed. Let $x \in X$ and let U be a neighbourhood of the fibre $p^{-1}(x)$. Let $m = \delta(x)$ and let $\langle g, \boldsymbol{a}\rangle = x$. We may assume that $U \subset G \times (\Delta_n \setminus \Delta_{m-1})$. Observe that $p^{-1}(x) = gK_m \times \{\boldsymbol{a}\} = (\pi_m \times \mathrm{id})^{-1}(gK_m, \boldsymbol{a})$, so since $\pi_m \times \mathrm{id}$

is a proper and hence closed map, Lemma 1.10.1 shows that there exists a neighbourhood W of $(gK_m, \boldsymbol{a})$ such that $(\pi_m \times \mathrm{id})^{-1}(W) \subset U$. But then $(\pi_m \times \mathrm{id})^{-1}(W) = (\omega_m \circ p_m)^{-1}(W)$ and hence $p^{-1}({\omega_m}^{-1}(W))$ is a saturated neighbourhood of the fibre $p^{-1}(x)$ which is contained in U. Since p is a quotient map by definition, p is a closed map by Lemma 1.10.2.

7. Let $\langle g, \boldsymbol{a}\rangle \neq \langle g', \boldsymbol{a}'\rangle \in X$. If $\boldsymbol{a} \neq \boldsymbol{a}'$, then $\boldsymbol{a}$ and $\boldsymbol{a}'$ have disjoint neighbourhoods U and U' in Δ_n. Then $\varpi^{-1}(U)$ and $\varpi^{-1}(U')$ are disjoint neighbourhoods of $\langle g, \boldsymbol{a}\rangle$ and $\langle g', \boldsymbol{a}'\rangle$ in X.

   Suppose then that $\boldsymbol{a} = \boldsymbol{a}'$. It follows that $gK_{\delta(\boldsymbol{a})} \neq g'K_{\delta(\boldsymbol{a})}$. Since $K_{\delta(\boldsymbol{a})}$ is closed in G, $G/K_{\delta(\boldsymbol{a})}$ is Hausdorff by Proposition 1.18. Therefore, $gK_{\delta(\boldsymbol{a})}$ and $g'K_{\delta(\boldsymbol{a})}$ have disjoint neighbourhoods U and U' in $G/K_{\delta(\boldsymbol{a})}$. Then ${\gamma_{\delta(\boldsymbol{a})}}^{-1}(U)$ and ${\gamma_{\delta(\boldsymbol{a})}}^{-1}(U')$ are disjoint neighbourhoods of $\langle g, \boldsymbol{a}\rangle$ and $\langle g', \boldsymbol{a}'\rangle$ in $X|(\Delta_n \setminus \Delta_{\delta(\boldsymbol{a})-1})$ and hence also in X.
8. Recall that $G \times \Delta_n$ is a proper left G-space, see Examples 1.14 and 1.15. Since $p\colon G \times \Delta_n \to X$ is a surjective and proper left G-map, it follows from Proposition 1.17.1 and 1.17.2 that X is a proper left G-space.
9. Proceed as in 8. Observe that if $K_0, \dots, K_n$ are closed in G, then $N(K_0), \dots,$ $N(K_n)$ are closed in G by Proposition 1.18. Hence $N_0^n \subset G$ is closed. Therefore, $G \times \Delta_n$ is a proper right N_0^n-space, see Examples 1.14, 1.15 and 1.16.

   □

The following lemma provides a very efficient tool for analysing the continuity of a map with values in an equivariant simplex.

Lemma 2.2. *Let X be an equivariant simplex as above and let Y be any topological space. A function $f\colon Y \to X$ is continuous if*

1. *$\varpi \circ f\colon Y \to \Delta_n$ is continuous; and*
2. *$\gamma_m \circ f\colon (\varpi \circ f)^{-1}(\Delta_n \setminus \Delta_{m-1}) \to G/K_m$ is continuous for all $0 \leq m \leq n$.*

Proof. Let $y \in Y$ and let U be a neighbourhood of $f(y)$. Let $m = \delta((\varpi \circ f)(y))$. By virtue of Theorem 2.1.5 there exists a neighbourhood W of $\omega_m(f(y))$ such that ${\omega_m}^{-1}(W) \subset U$. Since $\varpi \circ f$ is continuous, $(\varpi \circ f)^{-1}(\Delta_n \setminus \Delta_{m-1})$ is a neighbourhood of y. Since also $\gamma_m \circ f$ is continuous,

$$\omega_m \circ f\colon (\varpi \circ f)^{-1}(\Delta_n \setminus \Delta_{m-1}) \to G/K_m \times (\Delta_n \setminus \Delta_{m-1})$$

is continuous. Hence $(\varpi \circ f)^{-1}(\Delta_n \setminus \Delta_{m-1})$ contains a neighbourhood V of y such that $(\omega_m \circ f)(V) \subset W$. But then

$$f(V) \subset {\omega_m}^{-1}(W) \subset U,$$

which shows that f is continuous. □

Let us finish this section by recording an obvious result.

Proposition 2.3. *Let $x \in X, g \in G$ and $\boldsymbol{a} \in \Delta_n$. Then $x = \langle g, \boldsymbol{a}\rangle$ if and only if*

1. *$\varpi(x) = \boldsymbol{a}$; and*
2. *$\gamma_{\delta(\boldsymbol{a})}(x) = gK_{\delta(\boldsymbol{a})}$.* □

2.3. Construction of equivariant simplexes using cones and twisted products

In Section 2.2 the equivariant simplex $X = \Delta_n(G; K_0, \dots, K_n)$ was defined by making certain identifications on the set $G \times \Delta_n$. In this section we show that X is G-homeomorphic to a space constructed by a repeated application of the cone and twisted product functors starting from the one point space. In Section 2.4 we use this information to show that X is G-invariantly metrisable whenever G is metrisable.

Proposition 2.4. *There exists a K_0-homeomorphism*

$$h\colon C(\Delta_{n-1}(K_0; K_1, \dots, K_n)) \to \Delta_n(K_0; K_0, \dots, K_n).$$

Proof. We identify $C(\Delta_{n-1})$ and Δ_n with the aid of the proper map $q\colon I \times \Delta_{n-1} \to \Delta_n$, $(t, \boldsymbol{a}) \mapsto (1-t, ta_0, \dots, ta_{n-1})$, and this gives meaning to $[t, \boldsymbol{a}]$.

Let us consider the K_0-space $I \times K_0 \times \Delta_{n-1}$, $(k', (t, k, \boldsymbol{a})) \mapsto (t, k'k, \boldsymbol{a})$. Consider the K_0-maps

$$\begin{array}{rccc} f\colon & I \times K_0 \times \Delta_{n-1} & \to & C(\Delta_{n-1}(K_0; K_1, \dots, K_n)) \\ & (t, k, \boldsymbol{a}) & \mapsto & [t, \langle k, \boldsymbol{a} \rangle] \end{array}$$

and

$$\begin{array}{rccc} f'\colon & I \times K_0 \times \Delta_{n-1} & \to & \Delta_n(K_0; K_0, \dots, K_n) \\ & (t, k, \boldsymbol{a}) & \mapsto & \langle k, [t, \boldsymbol{a}] \rangle \end{array}.$$

Note that $f = q \circ (\mathrm{id} \times p)$ is a quotient map since p is a proper map by Theorem 2.1.6 and q is a quotient map by definition. Similarly, $f' = p \circ (\mathrm{id} \times q) \circ \sigma$ is a quotient map since $\sigma(t, k, \boldsymbol{a}) = (k, t, \boldsymbol{a})$ is a homeomorphism, q is a proper map and p is a quotient map by definition.

Let $(t, k, \boldsymbol{a})$ and $(t', k', \boldsymbol{a}')$ be two points in $I \times K_0 \times \Delta_{n-1}$. Observe that $[t, \langle k, \boldsymbol{a} \rangle] = [t', \langle k', \boldsymbol{a}' \rangle]$ if and only if $t = t'$ and $\langle k, \boldsymbol{a} \rangle = \langle k', \boldsymbol{a}' \rangle$, or $t = 0 = t'$. This again holds if and only if $t = t'$, $\boldsymbol{a} = \boldsymbol{a}'$ and $k'^{-1}k \in K_{\delta(\boldsymbol{a})}$, or $t = 0 = t'$.

Similarly, $\langle k, [t, \boldsymbol{a}] \rangle = \langle k', [t', \boldsymbol{a}'] \rangle$ if and only if $[t, \boldsymbol{a}] = [t', \boldsymbol{a}']$ and $k'^{-1}k \in K_{\delta([t,\boldsymbol{a}])}$, which again holds if and only if $t = t'$, $\boldsymbol{a} = \boldsymbol{a}'$ and $k'^{-1}k \in K_{\delta(\boldsymbol{a})}$, or $t = 0 = t'$. It follows that f and f' have identical fibres and hence there exists a unique K_0-homeomorphism h satisfying $f' = hf$, see Example 1.6. □

Proposition 2.5. *There exists a G-homeomorphism*

$$h\colon G \times_{K_0} \Delta_n(K_0; K_0, \dots, K_n) \to \Delta_n(G; K_0, \dots, K_n).$$

Proof. Let us consider the G-space $G \times K_0 \times \Delta_n$, $(g', (g, k, \boldsymbol{a})) \mapsto (g'g, k, \boldsymbol{a})$. Consider the G-maps

$$\begin{array}{rccc} f\colon & G \times K_0 \times \Delta_n & \to & G \times_{K_0} \Delta_n(K_0; K_0, \dots, K_n) \\ & (g, k, \boldsymbol{a}) & \mapsto & [g, \langle k, \boldsymbol{a} \rangle] \end{array}$$

and

$$\begin{array}{rccc} f'\colon & G \times K_0 \times \Delta_n & \to & \Delta_n(G; K_0, \dots, K_n) \\ & (g, k, \boldsymbol{a}) & \mapsto & \langle gk, \boldsymbol{a} \rangle \end{array}.$$

Note that $f = q \circ (\mathrm{id} \times p)$ is a quotient map since p is a proper map by Theorem 2.1.6 and q is a quotient map by definition. Similarly, $f' = p \circ (\mu \times \mathrm{id})$ is a quotient map since $\mu\colon G \times K_0 \to G, (g, k) \mapsto gk$, is a proper map by Proposition 1.11 and p is a quotient map by definition.

Let $(g, k, \boldsymbol{a})$ and $(g', k', \boldsymbol{a}')$ be two points in $G \times K_0 \times \Delta_n$. Observe that $[g, \langle k, \boldsymbol{a} \rangle] = [g', \langle k', \boldsymbol{a}' \rangle]$ if and only if ${g'}^{-1} g \in K_0$ and $\langle k', \boldsymbol{a}' \rangle = {g'}^{-1} g \langle k, \boldsymbol{a} \rangle$, which again holds if and only if $\boldsymbol{a} = \boldsymbol{a}'$ and ${k'}^{-1} {g'}^{-1} gk \in K_{\delta(\boldsymbol{a})}$.

Similarly, $\langle gk, \boldsymbol{a} \rangle = \langle g'k', \boldsymbol{a}' \rangle$ if and only if $\boldsymbol{a} = \boldsymbol{a}'$ and $(g'k')^{-1}(gk) \in K_{\delta(\boldsymbol{a})}$. It follows that f and f' have identical fibres and hence there exists a unique G-homeomorphism h satisfying $f' = hf$, see Example 1.6. □

Let G be a topological group and let $K_0 \supset \cdots \supset K_n$ be compact subgroups. It follows from Propositions 2.4 and 2.5 that $\Delta_n(G; K_0, \dots, K_n)$ is G-homeomorphic to a space constructed by a repeated application of the cone and twisted product functors starting from the 0-dimensional simplex $\Delta_0(K_n; K_n)$ which is easily identified as the one point K_n-space K_n/K_n.

2.4. Construction of a G-invariant metric on the equivariant simplex

We say that a G-space X is *G-invariantly metrisable* if the topology of X is induced by a metric d that satisfies $d(gx, gy) = d(x, y)$ for all g, x and y. In this section we give a direct proof that the equivariant simplex $X = \Delta_n(G; K_0, \dots, K_n)$ is G-invariantly metrisable whenever G is a *metrisable group*, i.e., the topology of G is induced by a metric. Our approach makes use of a construction of an equivariant simplex by means of cones and twisted products starting from the one point space. The existence of such a construction follows by Section 2.3.

The following theorems are formulated using pseudo-metrics rather than metrics. This has the advantage that now there is no need to check explicitly that our constructions yield metrics at each step. Recall that a pseudo-metric is a metric if the topology that it induces is T_1; in particular, if the topology is Hausdorff, see [Kel55], p. 124.

Theorem 2.6. *Let G be a pseudo-metrisable group and let $K \subset G$ be a compact subgroup. Then G admits a pseudo-metric d that is both left G-invariant and right K-invariant.*

Proof. In [HR63], Theorem 8.6 it is shown that a *compact* pseudo-metrisable group admits a pseudo-metric that is both left and right invariant. The proof below is essentially the same.

First, G admits a left G-invariant pseudo-metric σ, see [HR63], Theorem 8.3. Then define

$$\rho(x,y) = \sup\{\ \sigma(xk,yk) \mid k \in K\ \},$$

which is finite since a continuous real-valued function attains its maximum in a compact set. Now ρ is a pseudo-metric on G which is both left G-invariant and right K-invariant. It remains to check that ρ is topologically equivalent to σ.

It suffices to consider balls at e. Let $r > 0$. Clearly, $B_\rho(e,r) \subset B_\sigma(e,r)$. For the opposite direction, let us first consider the continuous function

$$\begin{array}{rccc} f\colon & K \times G & \to & G \\ & (k,g) & \mapsto & k^{-1}gk \end{array}.$$

Since the set $K \times \{e\}$ maps into $B_\sigma(e,r/2)$ and since K is compact, it follows from Lemma 1.12 that there exists an $s > 0$ such that $f(K \times B_\sigma(e,s)) \subset B_\sigma(e,r/2)$. Hence, for all $k \in K$ and all $y \in B_\sigma(e,s)$ we have $\sigma(ek,yk) = \sigma(e,k^{-1}yk) < r/2$, which implies that $\rho(e,x) < r$ and hence $B_\sigma(e,s) \subset B_\rho(e,r)$. □

Proposition 2.7. *Let G be a topological group and let X be a compact G-space that is G-invariantly pseudo-metrisable. Then the cone $C(X)$ is G-invariantly pseudo-metrisable.*

Proof. Since X is G-invariantly pseudo-metrisable, by definition there exists a G-invariant pseudo-metric d which induces the topology of X. Since X is compact, we may well assume that $d(X) \leq 1$.

Recall from Section 1.4 that $q\colon I \times X \to C(X)$ is a proper map. We define a real valued function

$$\begin{array}{rccc} e\colon & (I \times X) \times (I \times X) & \to & [0,\infty[\\ & ((r,x),(t,z)) & \mapsto & \min\{r,t\}d(x,z) + |r-t|. \end{array}$$

This function is symmetric and G-invariant. To see that e factors through $C(X) \times C(X)$ we need only to note that if $r = 0$, then $e((r,x),(t,z)) = |t|$ and similarly if $t = 0$, then $e((r,x),(t,z)) = |r|$. Since $q \times q$ is a quotient map, the induced map e' is continuous. To see that e' is a pseudo-metric, it remains to show that e satisfies the triangle inequality.

Let $(y,s) \in I \times X$. If $s \geq \min\{r,t\}$, then

$$\begin{aligned} \min\{r,t\}d(x,z) + |r-t| &\leq \min\{r,t\}(d(x,y) + d(y,z)) + |r-s| + |s-t| \\ &\leq \min\{r,s\}d(x,y) + \min\{s,t\}d(y,z) + |r-s| + |s-t|. \end{aligned}$$

Suppose then that $s \leq \min\{r,t\}$. By symmetry, we may suppose that $\min\{r,t\} = r$. Then $r = s + |r-s|$. The calculation

$$\begin{aligned} \min\{r,t\}d(x,z) + |r-t| &= \overset{\leq d(x,y)+d(y,z)}{sd(x,z)} + |r-s|\overset{\leq 1}{d(x,z)} + \overset{\leq |s-t|}{|r-t|} \\ &\leq \min\{r,s\}d(x,y) + \min\{s,t\}d(y,z) + |r-s| + |s-t| \end{aligned}$$

shows that e satisfies the triangle inequality. The same also holds for e'.

Since X is compact, if follows from Lemma 1.12 that the point $[0,x]$ has a neighbourhood basis consisting of sets of the form $q([0,t[\times X)$ where $t>0$. Since $e'([0,x],[t,z]) = |t|$, the sets $q([0,t[\times X)$ where $t>0$ coincide with the e'-balls $B_{e'}([0,x],t)$. That e' gives the topology of $C(X)$ at all other points as well is easy to see. □

Theorem 2.8. *Let X be a G-space with a G-invariant pseudo-metric d. Then*

$$\begin{aligned} d'\colon\quad X/G\times X/G &\to [0,\infty[\\ (Gx,Gy) &\mapsto \inf\{\, d(x,gy) \mid g\in G\,\} \end{aligned}$$

is a pseudo-metric on X/G which induces the topology of X/G.

Proof. It is routine to check that d' is a well-defined pseudo-metric on X/G. For the sake of completeness, let us present the triangle inequality. Let $x,y,z\in X$. Given $\epsilon>0$, find g and g' such that $d(x,gy)<d'(Gx,Gy)+\epsilon/2$ and $d(y,g'z)<d'(Gy,Gz)+\epsilon/2$. By invariance, $d(y,g'z)=d(gy,gg'z)$, and hence

$$d(x,gg'z)\le d(x,gy)+d(gy,gg'z)<d'(Gx,Gy)+d'(Gy,Gz)+\epsilon.$$

It follows that

$$d'(Gx,Gz)<d'(Gx,Gy)+d'(Gy,Gz)+\epsilon.$$

This holds for any $\epsilon>0$ and so the triangle inequality follows.

Having shown that d' is a pseudo-metric on the set X/G, we show that it induces the quotient topology on X/G. The orbit map $\pi\colon X\to (X/G,d')$ is continuous since it does not increase distances, that is, it is 1-Lipschitz. It remains to show that $\pi^{-1}(U)$ is open in X only if U is open in $(X/G,d')$. Let $x\in X$ be such that $Gx\in U$. Since $\pi^{-1}(U)$ is open in X, there exists an $r>0$ such that $B_d(x,r)\subset\pi^{-1}(U)$. Now suppose that $d'(Gx,Gy)<r$. Then $d(x,gy)<r$ for some $g\in G$. It follows that $gy\in B_d(x,r)$, and hence $gy\in\pi^{-1}(U)$, and therefore $Gy\in U$. This shows that U is open in the pseudo-metric topology induced by d'. □

Proposition 2.9. *Let G be a pseudo-metrisable group, let $K\subset G$ be a compact subgroup and let X be a K-space that is K-invariantly pseudo-metrisable. Then the twisted product $G\times_K X$ is G-invariantly pseudo-metrisable.*

Proof. Let e_X be a K-invariant pseudo-metric on X and let e_G be a pseudo-metric on G that is both left G-invariant and right K-invariant, see Theorem 2.6. Then

$$d((g_1,x_1),(g_2,x_2))=e_G(g_1,g_2)+e_X(x_1,x_2)$$

defines a pseudo-metric on $G\times X$ that is both G-invariant under $g(g_1,x_1)=(gg_1,x_1)$ and K-invariant under $k(g_1,x_1)=(g_1k^{-1},kx_1)$. The construction in Theorem 2.8 gives a pseudo-metric on $G\times_K X$ which is evidently G-invariant. □

Lemma 2.10. *If $f\colon X\to Y$ is a G-homeomorphism of G-spaces and if the topology of Y is induced by a G-invariant pseudo-metric d', then the topology of X is induced by the G-invariant pseudo-metric $d(x,y)=d'(f(x),f(y))$.* □

Theorem 2.11. *Let G be a metrisable group, let $K_0 \supset \cdots \supset K_n$ be compact subgroups and let X denote the equivariant simplex $\Delta_n(G; K_0, \ldots, K_n)$. Then X is G-invariantly metrisable.*

Proof. Since G is metrisable, the compact subgroups $K_0, \ldots, K_n$ are closed in G, and hence X is Hausdorff by Theorem 2.1.7. Therefore, it suffices to show that the topology of X is induced by a G-invariant pseudo-metric. It was noted in Lemma 2.10 that it suffices to show that X is G-homeomorphic to a G-invariantly metrisable G-space.

In Section 2.3 it was shown that X is G-homeomorphic to a space constructed by a repeated application of the cone and twisted product functors starting from the 0-dimensional simplex $\Delta_0(K_n; K_n)$ which is easily identified as the one point K_n-space K_n/K_n. With the possible exception of the last twisted product, all intermediate steps yield compact spaces. Therefore, it follows from Propositions 2.7 and 2.9 that X is G-homeomorphic to a G-invariantly pseudo-metrisable space. □

Remark 2.12. With the exception of Theorem 2.8, the results 2.6 – 2.9 are correct if every occurrence of words "pseudo-metric" and "pseudo-metrisable" is replaced by "metric" and "metrisable", respectively. In Theorem 2.8 one has to include additional assumptions: according to Proposition 1.13 and the previously cited remark in [Kel55], p. 124, it suffices to assume that the action is proper, for example.

Part II. The group $\mathbf{CS}(X_n)$ and its strong equivariant deformation retract $N_0^n \cdot \eta$

In Part I we showed that for any topological group G and any compact subgroups $K_0 \supset \cdots \supset K_n$ one can construct an equivariant simplex $X = \Delta_n(G; K_0, \ldots, K_n)$. It was shown that X is a proper G-space, the orbit space of which can be identified with the simplex Δ_n. Moreover, it was shown that X is G-invariantly metrisable whenever G is a metrisable group. We take this as our standing assumption here in Part II.

Here in Part II we work with function spaces that consist of cross-sections of the *left* G-space X. Throughout this thesis function spaces carry the *compact-open* topology. The required information about the compact-open topology is recalled in Section 3. Since the orbit space of X is Δ_n, constructions of deformations of function spaces may often be carried out using the familiar results that are found in Propositions 3.4 and 3.3.

In Section 4.1 we show that the set $\mathbf{CS}(X)$ of all cross-sections of X is a G-invariantly metrisable proper G-space. However, our interest does not lie so much in $\mathbf{CS}(X)$ as it does in cross-sections of certain important N_0^n-invariant subsets

$$X_0 \subset \cdots \subset X_n$$

of X, which are defined in Section 4.2. These induce an increasing sequence

$$\mathbf{CS}(X_0) \subset \mathbf{CS}(X_1) \subset \cdots \subset \mathbf{CS}(X_n)$$

of N_0^n-invariant subspaces of $\mathbf{CS}(X)$. In Section 4.3 we show that $\mathbf{CS}(X_n)$ is a topological group with a neutral element η and the sets $\mathbf{CS}(X_k)$ are its subgroups. In Section 4.4 we show that the action of N_0^n on $\mathbf{CS}(X_k)$ has the group K_n as its ineffective kernel and that it commutes with the group law implying that the orbit $N_0^n \cdot \eta$ is a subgroup of $\mathbf{CS}(X_k)$. Moreover, the effective action of N_0^n/K_n on $\mathbf{CS}(X_k)$ can be identified as the action of the subgroup $N_0^n \cdot \eta$. Our main goal in Part II is to give sufficient conditions under which *the subgroup $N_0^n \cdot \eta$ is a strong equivariant deformation retract of* $\mathbf{CS}(X_n)$.

In Section 5 we discuss on a general level the conditions and implications of the case when a subgroup M is an M-equivariant (neighbourhood) deformation retract of a topological group N. Theorem 5.1 provides a satisfactory characterisation of this situation by virtue of the existence of local cross-sections and local contraction properties.

In Section 6 we show that the N_0^n-space $\mathbf{CS}(X_0)$ is a strong N_0^n-deformation retract of $\mathbf{CS}(X_n)$. A key role is played by Proposition 4.4, which shows how certain types of parametrised deformations of the simplex Δ_n, the domain of the cross-sections, induce deformations of the spaces of cross-sections. However, it would be all too easy if Proposition 4.4 would solve our problem. The main obstacle encountered in this approach to constructing deformations of spaces of cross-sections is that the deformation does not allow the desired "dimension drop".

To circumvent the dimension drop problem we have to work on the range side of cross-sections. There is an intricate interplay of topology and algebra which makes it possible to define a set W_k such that $X_{k-1} \subset W_k \subset X_k$ and such that there exists an N_0^n-map $H_k\colon I \times W_k \to X_k$ over Δ_n which deforms W_k into X_{k-1}, see Theorem 6.4. The set W_k has the feature that there exists an N_0^n-map $\hat{R}_k\colon I \times \mathbf{CS}(X_k) \to \mathbf{CS}(X_k)$ which deforms $\mathbf{CS}(X_k)$ into $\mathbf{CS}(W_k)$, see Proposition 6.8. The construction of $\hat{R}_k$ makes use of a well-known theorem of C. H. Dowker on semicontinuous functions, see Lemma 6.6. The dimension drop is then handled by deforming the range W_k into X_{k-1}. This leads inductively to an N_0^n-deformation of $\mathbf{CS}(X_n)$ into $\mathbf{CS}(X_0)$, see Corollary 6.10.

In Section 7 we again apply Proposition 4.4, which now provides us an easy N_0^n-deformation of $\mathbf{CS}(X_0)$ onto the orbit $N_0^n \cdot \eta$, see Theorem 7.1. The main result of this thesis appears for the first time in Theorem 7.2 in the general form. It immediately yields a simplified version for Lie groups, see Theorem 7.3.

3. Basic properties of the compact-open topology

The basic properties of the compact-open topology are readily found in the literature. The purpose of this section is to introduce only those results that are needed later.

The set $\mathrm{C}(X, Y)$ of continuous maps from X to Y with the *compact-open topology* is denoted by $\mathbf{map}(X, Y)$. This is the topology generated by the *elementary neighbourhoods*

$$\mathrm{W}(A,B) = \{\ f\colon X \to Y \mid f(A) \subset B\ \}$$

where $A \subset X$ is compact and $B \subset Y$ is open. This means that finite intersections of elementary neighbourhoods form a basis for the compact-open topology.

3.1. Induced maps

Let X, Y and Z be topological spaces and let $h\colon Y \to Z$ be a continuous map. Then h induces a map

$$\begin{array}{rccc} h_*\colon & \mathbf{map}(X,Y) & \to & \mathbf{map}(X,Z) \\ & f & \mapsto & hf \end{array},$$

which we call *postcomposition* with h. For the proof of the following basic result, see [Dug66], XII.2.1(2).

Proposition 3.1. *The postcomposition map h_* induced by $h\colon Y \to Z$ is continuous.* □

Similarly, any continuous map $f\colon X \to Y$ induces a map

$$\begin{array}{rccc} f^*\colon & \mathbf{map}(Y,Z) & \to & \mathbf{map}(X,Z) \\ & h & \mapsto & hf \end{array},$$

which we call *precomposition* with f. For the proof of the following basic results, see [Dug66], XII.2.1(1) and [Jam84], Proposition 2.86.

Proposition 3.2. *The precomposition map f^* induced by $f\colon X \to Y$ is continuous. If f is a surjective proper map, then f^* is an embedding.* □

3.2. The evaluation map

The evaluation function

$$\begin{array}{rccc} \mathrm{e}\colon & \mathbf{map}(X,Y) \times X & \to & Y \\ & (f,x) & \mapsto & f(x) \end{array}$$

is a well-defined function. The following continuity statement is well-known, see [Dug66], Theorem XII.2.4(2), for example.

Proposition 3.3. *Let X be a locally compact Hausdorff space and let Y be any topological space. Then the evaluation function*

$$\mathrm{e}\colon \mathbf{map}(X,Y) \times X \to Y$$

is continuous. □

3.3. Adjoint maps

The exponential law of set theory gives a bijective correspondence between $Z^{X\times Y}$ and $(Z^Y)^X$. This correspondence adjusts quite naturally to continuous functions as we will now recall.

Let $f\colon X\times Y\to Z$ be a continuous map. For each $x\in X$ we obtain a continuous function

$$\begin{array}{rccc} f_x\colon & Y & \to & Z \\ & y & \mapsto & f(x,y) \end{array}.$$

The function $\hat{f}$ that assigns to each x the map f_x is called the *adjoint* of f. Once again, the following result is found in [Dug66], Theorem XII.3.1(1), for example.

Proposition 3.4. *Let $f\colon X\times Y\to Z$ be a continuous map. Then*

$$\begin{array}{rccc} \hat{f}\colon & X & \to & \mathbf{map}(Y,Z) \\ & x & \mapsto & f_x \end{array}$$

is continuous. □

3.4. The diagonal map

Let X be any topological space. Consider the *diagonal function*

$$\begin{array}{rccc} \Delta\colon & \mathrm{C}(\Delta_n,X)\times\mathrm{C}(\Delta_n,X) & \to & \mathrm{C}(\Delta_n,X\times X) \\ & (\alpha,\beta) & \mapsto & \Delta(\alpha,\beta) \end{array},$$

where $\Delta(\alpha,\beta)$ denotes the map $\boldsymbol{a}\mapsto(\alpha(\boldsymbol{a}),\beta(\boldsymbol{a}))$.

Proposition 3.5. *The diagonal map*

$$\begin{array}{rccc} \Delta\colon & \mathbf{map}(\Delta_n,X)\times\mathbf{map}(\Delta_n,X) & \to & \mathbf{map}(\Delta_n,X\times X) \\ & (\alpha,\beta) & \mapsto & \Delta(\alpha,\beta) \end{array}$$

is continuous for all spaces X.

Proof. Observe that the inverse of Δ is $(\mathrm{pr}_{1*},\mathrm{pr}_{2*})$. Since Δ_n is regular, it follows from Proposition 2.93 in [Jam84] that $(\mathrm{pr}_{1*},\mathrm{pr}_{2*})$ is a homeomorphism. Therefore, Δ is continuous. □

3.5. The product functor

The product functor is functor with two arguments. We confine the discussion to a partial functor obtained by fixing a space.

Proposition 3.6. *Let G be a topological group. Then the product functor*

$$\begin{array}{rccc} \mathrm{id}\times\bullet\colon & \mathbf{map}(\Delta_n,X) & \to & \mathbf{map}(G\times\Delta_n,G\times X) \\ & f & \mapsto & \mathrm{id}\times f \end{array}$$

is continuous.

Proof. Every topological group is regular, see [HR63], Corollary 4.7 and Theorem 4.2. Here by regular we mean that every neighbourhood of any given point contains a closed neighbourhood. Since G and Δ_n are regular spaces, the product functor

$$\mathbf{map}(G, G) \times \mathbf{map}(\Delta_n, X) \to \mathbf{map}(G \times \Delta_n, G \times X)$$

is continuous, see [Jam84], Corollary 2.96, for example. In particular, by fixing the first variable as id, the function $\mathrm{id} \times \bullet$ is continuous. □

4. Cross-sections of equivariant simplexes

Throughout Section 4 we assume that G is a metrisable group, $K_0 \supset \cdots \supset K_n$ are compact subgroups and X denotes the equivariant simplex $\Delta_n(G; K_0, \ldots, K_n)$, which we consider as a left G-space.

4.1. The G-space $\mathbf{CS}(X)$

By a cross-section of an equivariant simplex X we mean a continuous map $\alpha\colon \Delta_n \to X$ which satisfies the condition that $\varpi\alpha = \mathrm{id}$. The set of all cross-sections of X equipped with the compact-open topology is denoted by $\mathbf{CS}(X)$. The neutral cross-section η, see Section 2.2, will receive special attention from here on.

The equivariant simplex X has a continuous G-action $\Phi\colon G \times X \to X$. The adjoint $\varphi = \hat{\Phi}$ gives a canonical homomorphism

$$\begin{array}{rccc} \varphi\colon & G & \to & \mathbf{Homeo}_{\Delta_n}(X) \\ & g & \mapsto & \varphi(g) \end{array},$$

where $\varphi(g)$ is the transformation $x \mapsto gx$. Here $\mathbf{Homeo}_{\Delta_n}(X)$ denotes the group of homeomorphisms $f\colon X \to X$ *over* Δ_n, that is, with $\varpi f = \varpi$. For any $g \in G$ and $\alpha \in \mathbf{CS}(X)$ the transformation $\varphi(g)$ is a map over Δ_n, and hence $\varphi(g)\alpha$ is a global cross-section. Since φ is a homomorphism, we obtain a left action

$$\begin{array}{rccc} \Phi\colon & G \times \mathbf{CS}(X) & \to & \mathbf{CS}(X) \\ & (g, \alpha) & \mapsto & \varphi(g)\alpha \end{array}.$$

To simplify the notation, we denote $\varphi(g)\alpha$ by $g \cdot \alpha$. This action is continuous as we will next explain.

Observe that there is a function

$$\begin{array}{rccc} \check{\Phi}\colon & G \times \mathbf{CS}(X) \times \Delta_n & \to & X \\ & (g, \alpha, \boldsymbol{a}) & \mapsto & g\,\mathrm{e}(\alpha, \boldsymbol{a}) \end{array},$$

which is continuous by Proposition 3.3 and by the fact that $\Phi\colon G \times X \to X$ is continuous. Since $\Phi\colon G \times \mathbf{CS}(X) \to \mathbf{CS}(X)$ is the adjoint of $\check{\Phi}$, it is continuous by Proposition 3.4.

Proposition 4.1. $\mathbf{CS}(X)$ *is a G-invariantly metrisable proper G-space. The orbital map*

$$\begin{array}{rccc} \lambda\colon & G & \to & \mathbf{CS}(X) \\ & g & \mapsto & g\cdot\eta \end{array}$$

is a proper G-map.

Proof. Since G is a metrisable group, X has a left G-invariant metric d by Theorem 2.11. Since Δ_n is compact, the topology of $\mathbf{CS}(X)$ is given by the metric

$$d^+(\alpha,\beta) = \sup\{\ d(\alpha(\boldsymbol{a}),\beta(\boldsymbol{a})) \mid \boldsymbol{a}\in\Delta_n\ \},$$

which is shown in [Dug66], XII.8.2(3). This metric is G-invariant.

Note that in particular $\mathbf{CS}(X)$ is a Hausdorff G-space. Let $\boldsymbol{a}\in\Delta_n$ be any point. In the compact-open topology, evaluation at *one* point is certainly continuous. Hence the map

$$\begin{array}{rccc} f\colon & \mathbf{CS}(X) & \to & X \\ & \alpha & \mapsto & \alpha(\boldsymbol{a}) \end{array}$$

is a continuous G-map. Since X is a proper G-space by Theorem 2.1.8, also $\mathbf{CS}(X)$ is a proper G-space by Proposition 1.20.1.

That λ is a proper G-map follows then by Proposition 1.13. □

4.2. The N_0^n-spaces X^k and X_k

Recall that in Section 2.2 we agreed that N_0^n denotes the subgroup $N(K_0)\cap\cdots\cap N(K_n)$ of G. Let us now generalise this notation. For any $0\le k\le n$ we denote the group $N(K_k)$ by N_k and the group $N(K_k)\cap\cdots\cap N(K_n)$ by N_k^n.

Let $k\le i\le n$. Since K_i is compact, the canonical map $\pi_i\colon G\to G/K_i$ is proper, see Proposition 1.11. Since G is metrisable, the groups K_i are closed and hence so is N_k^n, by Proposition 1.18. We denote by N_k^n/K_i the image of N_k^n under π_i. The restricted map $\pi_i\colon N_k^n\to N_k^n/K_i$ is proper by Proposition 1.9.1; in particular, the restricted map is a quotient map. Since N_k^n is a subgroup of N_i and since K_i is a normal subgroup of N_i, it follows that N_k^n/K_i has a natural structure of a topological group.

We will now define certain N_0^n-*invariant* subsets of the *left G-space* X. We are interested in the N_0^n-invariant subsets

$$X_0\subset X_1\subset\cdots\subset X_n,$$

where X_k $(0\le k\le n)$ denotes the subset of X that consists of points x such that

$$\gamma_{\delta(x)}(x)\in N_{\delta(x)}^n/K_{\delta(x)}$$

when $\delta(x)\le k$, and

$$\gamma_{\delta(x)}(x)\in N_k^n/K_{\delta(x)}$$

when $\delta(x)>k$. Letting $m=\min\{\delta(x),k\}$, the above condition can be expressed as $\gamma_{\delta(x)}(x)\in N_m^n/K_{\delta(x)}$, which we sometimes find convenient. The spaces X_k should not be confused with lower dimensional equivariant simplexes since $\varpi(X_k) = \Delta_n$ for all k.

In Section 4.1 we introduced the G-space $\mathbf{CS}(X)$. The notion of a cross-section of X is tied to the projection $\varpi\colon X \to \Delta_n$. Therefore, since the spaces $X_k \subset X$ are such that $\varpi(X_k) = \Delta_n$, it makes perfect sense to speak of maps $\alpha\colon \Delta_n \to X_k$ as cross-sections of X_k if $\varpi\alpha = \mathrm{id}$. Such cross-sections always exist since the neutral cross-section η belongs to every $\mathbf{CS}(X_k)$. As the spaces X_k form an increasing sequence of N_0^n-invariant subsets of X, the spaces $\mathbf{CS}(X_k)$ also are arranged in an increasing sequence

$$\mathbf{CS}(X_0) \subset \mathbf{CS}(X_1) \subset \cdots \subset \mathbf{CS}(X_n)$$

of N_0^n-invariant subspaces of $\mathbf{CS}(X)$.

The space

$$X_n = \{\, x \in X \mid \gamma_{\delta(x)}(x) \in N^n_{\delta(x)}/K_{\delta(x)} \,\}$$

has special importance: in Section 4.3 we will show that $\mathbf{CS}(X_n)$ has the structure of a topological group and later we will see that $\mathbf{CS}(X_n)$ is isomorphic to $\mathbf{Homeo}^G_{\Delta_n}(X)$, the group of G-equivariant homeomorphisms $f\colon X \to X$ over Δ_n.

Perhaps the best way to highlight the sets X_k is to give the following simple result.

Proposition 4.2. *Let α be a cross-section of X_n. Then $\alpha(\Delta_k) \subset X_k$.*

Proof. Let $\boldsymbol{a} \in \Delta_k$. Since $\alpha(\boldsymbol{a}) \in X_n$, it follows that

$$\gamma_{\delta(\boldsymbol{a})}(\alpha(\boldsymbol{a})) \in N^n_{\delta(\boldsymbol{a})}/K_{\delta(\boldsymbol{a})}.$$

Since $\delta(\boldsymbol{a}) \le k$, the above criterion is sufficient to guarantee that $\alpha(\boldsymbol{a}) \in X_k$. □

The filtration by sets X_k $(0 \le k \le n)$ is used to construct a strong N_0^n-equivariant deformation retraction of $\mathbf{CS}(X_n)$ onto its subspace $\mathbf{CS}(X_0)$, see Section 6. It is then easy to find a strong N_0^n-equivariant deformation retraction of $\mathbf{CS}(X_0)$ onto the N_0^n-orbit of the neutral cross-section η, see Section 7.

For technical reasons we will also introduce certain N_0^n-invariant subsets X^k of the left G-space $G \times \Delta_n$. Recall that the map $p\colon G \times \Delta_n \to X$ is a proper map by Theorem 2.1.6, and hence so is the restricted N_0^n-map $p\colon N_k^n \times \Delta_n \to X$ for each $0 \le k \le n$ since N_k^n is closed in G. Since X_k is contained in the image of $p\colon N_k^n \times \Delta_n \to X$, we obtain yet another restricted N_0^n-map $p\colon p^{-1}(X_k) \to X_k$. This map is surjective and a proper map by Proposition 1.9.2. We denote $X^k = p^{-1}(X_k)$ and we stress that $X^k \subset N_k^n \times \Delta_n$. Note that

$$X^0 \subset X^1 \subset \cdots \subset X^n.$$

4.3. The topological groups $\mathbf{CS}(X_k)$

We now proceed to show that $\mathbf{CS}(X_n)$ has a natural group structure and that all $\mathbf{CS}(X_k)$, $0 \le k \le n$, are its subgroups. The required operations are very similar to those in standard factor groups.

Recall the definition of X_n and X^n and that these are spaces over Δ_n and recall that the restricted map $p\colon X^n \to X_n$ is a proper map. Consequently, the product map $p \times p\colon X^n \times X^n \to X_n \times X_n$ is also a proper map.

There is a well-defined involution

$$\begin{array}{cccc} \iota^n\colon & X^n & \to & X^n \\ & (g, \boldsymbol{a}) & \mapsto & (g^{-1}, \boldsymbol{a}) \end{array}$$

over Δ_n. Since ι^n is compatible with the fibres of the quotient map p, there exists an induced involution $\iota_n\colon X_n \to X_n$ over Δ_n.

Consider next the fibre products

$$X^n \times_{\mathrm{pr}} X^n = \{\ ((g, \boldsymbol{a}), (g', \boldsymbol{a})) \mid (g, \boldsymbol{a}), (g', \boldsymbol{a}) \in X^n\ \}$$

and

$$X_n \times_{\varpi} X_n = \{\ (\langle g, \boldsymbol{a}\rangle, \langle g', \boldsymbol{a}\rangle) \mid \langle g, \boldsymbol{a}\rangle, \langle g', \boldsymbol{a}\rangle \in X_n\ \}.$$

Note that

$$X^n \times_{\mathrm{pr}} X^n = (p \times p)^{-1}(X_n \times_{\varpi} X_n).$$

This is important since now it follows that the restricted map

$$p \times p\colon X^n \times_{\mathrm{pr}} X^n \to X_n \times_{\varpi} X_n$$

is also a proper map, see Proposition 1.9.2.

By strict analogy with how it can be proved that normalisers of subgroups are subgroups one can show that the map

$$\begin{array}{cccc} \mu^n\colon & X^n \times_{\mathrm{pr}} X^n & \to & X^n \\ & ((g, \boldsymbol{a}), (g', \boldsymbol{a})) & \mapsto & (gg', \boldsymbol{a}) \end{array}$$

is well-defined. By analogy with factor groups, it induces a map

$$\begin{array}{cccc} \mu_n\colon & X_n \times_{\varpi} X_n & \to & X_n \\ & (\langle g, \boldsymbol{a}\rangle, \langle g', \boldsymbol{a}\rangle) & \mapsto & \langle gg', \boldsymbol{a}\rangle \end{array}.$$

What we have constructed could be called a *group bundle*, a space over Δ_n with a fibrewise multiplication. It is not a group since in general there is no natural identity element. However, the space of cross-sections of X_n can now be turned into a topological group using pointwise operations.

Given $\alpha, \beta \in \mathbf{CS}(X_n)$, let us define the product $\alpha \star \beta$ as the unique cross-section in $\mathbf{CS}(X_n)$ that satisfies

$$(\alpha \star \beta)(\boldsymbol{a}) = \mu_n(\alpha(\boldsymbol{a}), \beta(\boldsymbol{a})).$$

In practice this means that one chooses for each $\boldsymbol{a} \in \Delta_n$ representatives $l, l' \in N^n_{\delta(\boldsymbol{a})}$ such that $\alpha(\boldsymbol{a}) = \langle l, \boldsymbol{a}\rangle$ and $\beta(\boldsymbol{a}) = \langle l', \boldsymbol{a}\rangle$, which implies that $(\alpha \star \beta)(\boldsymbol{a}) = \langle ll', \boldsymbol{a}\rangle$. For each $\alpha \in \mathbf{CS}(X_n)$ we also define the inverse α^{-1} as the cross-section

$$\alpha^{-1}(\boldsymbol{a}) = \iota_n(\alpha(\boldsymbol{a})).$$

Again, in practice this means that $\alpha^{-1}(\boldsymbol{a}) = \langle l^{-1}, \boldsymbol{a}\rangle$. Now $(\mathbf{CS}(X_n), \star)$ is a group. The neutral element of $\mathbf{CS}(X_n)$ is the neutral cross-section η.

Theorem 4.3. *The space of cross-sections* $\mathbf{CS}(X_n)$ *with pointwise operations is a topological group. Moreover, every* $\mathbf{CS}(X_k)$, $0 \le k \le n$, *is a closed subgroup of* $\mathbf{CS}(X_n)$.

Proof. It remains to show that the group operations are continuous. To see that the multiplication $\star$ is continuous it is convenient to write $\star$ as the function $\mu_{n*} \circ \Delta$, where Δ is the diagonal function

$$\begin{array}{rccc} \Delta\colon & \mathbf{CS}(X_n) \times \mathbf{CS}(X_n) & \to & \mathbf{map}(\Delta_n, X_n \times_{\varpi} X_n) \\ & (\alpha, \alpha') & \mapsto & \Delta(\alpha, \alpha'), \end{array}$$

with $\Delta(\alpha, \alpha')$ being the map $\boldsymbol{a} \mapsto (\alpha(\boldsymbol{a}), \alpha'(\boldsymbol{a}))$. The continuity of Δ follows by Proposition 3.5 and by the fact that $X_n \times_{\varpi} X_n$ is a subspace of $X_n \times X_n$. The continuity of μ_{n*} follows by Proposition 3.1. Therefore, the multiplication in $\mathbf{CS}(X_n)$ is continuous.

The inverse-element function in $\mathbf{CS}(X_n)$ is given by function ι_{n*}, where ι_n refers to the involution of X_n constructed above. Again ι_{n*} is continuous by Proposition 3.1.

Let $0 \le k \le n$. To see that $\mathbf{CS}(X_k)$ is a subgroup, we first note that for any $\boldsymbol{a} \in \Delta_n$, the function

$$\begin{array}{rccc} f_{\boldsymbol{a}}\colon & \mathbf{CS}(X_n) & \to & N^n_{\delta(\boldsymbol{a})}/K_{\delta(\boldsymbol{a})} \\ & \alpha & \mapsto & \gamma_{\delta(\boldsymbol{a})}(\alpha(\boldsymbol{a})) \end{array}$$

is a continuous homomorphism. Then it is clear that

$$\mathbf{CS}(X_k) = \bigcap_{\boldsymbol{a} \in \Delta_n} {f_{\boldsymbol{a}}}^{-1}(N^n_{\min\{\delta(\boldsymbol{a}),k\}}/K_{\delta(\boldsymbol{a})})$$

is a closed subgroup of $\mathbf{CS}(X_n)$. □

4.4. The N^n_0-spaces $\mathbf{CS}(X_k)$

We recall from Section 4.1 that $\mathbf{CS}(X)$ is a G-invariantly metrisable proper G-space and from Section 4.2 that the spaces $\mathbf{CS}(X_k)$ are N^n_0-invariant subspaces of $\mathbf{CS}(X)$. Thus the action of G on $\mathbf{CS}(X)$ induces on $\mathbf{CS}(X_k)$ a proper action Φ of the closed subgroup N^n_0, see Example 1.16 and Proposition 1.20.2. Obviously, the N^n_0-space $\mathbf{CS}(X_k)$ is N^n_0-invariantly metrisable.

In Section 4.3 we saw that $\mathbf{CS}(X_n)$ is a topological group with respect to pointwise operation $\star$. The action Φ of N^n_0 on $\mathbf{CS}(X_n)$ and the group law in $(\mathbf{CS}(X_n), \star)$ are tied together via the formula

$$g \cdot (\alpha \star \alpha') = (g \cdot \alpha) \star \alpha'.$$

For if $\boldsymbol{a} \in \Delta_n$ and if $l, l' \in N^n_{\delta(\boldsymbol{a})}$ are such that $\alpha(\boldsymbol{a}) = l\eta(\boldsymbol{a})$ and $\alpha'(\boldsymbol{a}) = l'\eta(\boldsymbol{a})$, then $(\alpha \star \alpha')(\boldsymbol{a}) = ll'\eta(\boldsymbol{a})$ and $gl\eta(\boldsymbol{a}) = (g \cdot \alpha)(\boldsymbol{a})$ and so

$$(g \cdot (\alpha \star \alpha'))(\boldsymbol{a}) = ((g \cdot \alpha) \star \alpha')(\boldsymbol{a}).$$

One consequence of this law is that we obtain a homomorphism

$$\begin{array}{rccc} \lambda\colon & N^n_0 & \longrightarrow & \mathbf{CS}(X_n) \\ & g & \longmapsto & g \cdot \eta \end{array}$$

with kernel K_n onto the N^n_0-orbit of the neutral cross-section η, which is in fact a closed subgroup of $(\mathbf{CS}(X_n), \star)$ and lies in $\mathbf{CS}(X_0)$. The calculation that confirms this is done as follows:

$$\begin{aligned} \lambda(g) \star \lambda(g') &= (g \cdot \eta) \star (g' \cdot \eta) \\ &= g \cdot (\eta \star (g' \cdot \eta)) \\ &= g \cdot (g' \cdot \eta) \\ &= (gg') \cdot \eta = \lambda(gg'). \end{aligned}$$

That $N^n_0 \cdot \eta$ is closed in $\mathbf{CS}(X_k)$ for all $0 \le k \le n$ follows from the fact that λ is a proper map by Proposition 4.1 and by the fact that N^n_0 is closed in G. Since the kernel of λ is K_n, it follows that $N^n_0 \cdot \eta$ can be identified with the group N^n_0/K_n.

Notice also that

$$g \cdot \alpha = g \cdot (\eta \star \alpha) = (g \cdot \eta) \star \alpha = \lambda(g) \star \alpha,$$

which shows that the action Φ is the same action as the action of N^n_0 on $\mathbf{CS}(X_k)$ by left multiplication through the homomorphism λ.

4.5. Deformations of $\mathbf{CS}(X_k)$ induced by argument homotopies

Consider a cross-section $\alpha \in \mathbf{CS}(X)$. The value that α takes at a point $\boldsymbol{a}'$ is completely determined by the coordinate $\gamma_{\delta(\boldsymbol{a}')}(\alpha(\boldsymbol{a}')) \in G/K_{\delta(\boldsymbol{a}')}$, see Proposition 2.3. Suppose that we want to construct a new cross-section which has "the same" coordinate value but at a different point $\boldsymbol{a}$. The following technical statement shows how this can be done and even in a continuous way.

Proposition 4.4. *Let* $0 \le k \le n$ *and let*

$$F\colon I \times \mathbf{CS}(X_k) \times \Delta_n \to \Delta_n$$

be a continuous N^n_0*-invariant map such that*

1. $\delta(F(t, \alpha, \boldsymbol{a})) \ge \delta(\boldsymbol{a})$ *for all* $\boldsymbol{a}$*; and*
2. $\delta(F(t, \alpha, \boldsymbol{a})) = \delta(\boldsymbol{a})$ *when* $\delta(\boldsymbol{a}) < k$.

Then

$$\begin{array}{rccc} Q\colon & I \times \mathbf{CS}(X_k) \times \Delta_n & \longrightarrow & X_k \\ & (t, \alpha, \boldsymbol{a}) & \longmapsto & e_{\delta(\boldsymbol{a})}(\gamma_{\delta(\boldsymbol{a})}(\alpha(F(t, \alpha, \boldsymbol{a}))), \boldsymbol{a}) \end{array}$$

is a well-defined continuous N_0^n-equivariant map over Δ_n. The adjoint of Q is an N_0^n-equivariant map

$$\hat{Q}\colon I \times \mathbf{CS}(X_k) \to \mathbf{CS}(X_k).$$

Proof. Let us first show that Q is a well-defined function. Let $(t, \alpha, \boldsymbol{a}) \in I \times \mathbf{CS}(X_k) \times \Delta_n$. Observe first that since α is a cross-section and since F does not drop dimension, $\delta(\alpha(F(t,\alpha,\boldsymbol{a}))) = \delta(F(t,\alpha,\boldsymbol{a})) \geq \delta(\boldsymbol{a})$. It follows that $\gamma_{\delta(\boldsymbol{a})}$ can be applied to $\alpha(F(t,\alpha,\boldsymbol{a}))$ and hence $e_{\delta(\boldsymbol{a})}(\gamma_{\delta(\boldsymbol{a})}(\alpha(F(t,\alpha,\boldsymbol{a}))), \boldsymbol{a})$ represents a well-defined element of X. Next we show that the values of Q reside in X_k.

Namely, since $\alpha \in \mathbf{CS}(X_k)$, it follows that

$$\gamma_{\delta(F(t,\alpha,\boldsymbol{a}))}(\alpha(F(t,\alpha,\boldsymbol{a}))) \in N_m^n / K_{\delta(F(t,\alpha,\boldsymbol{a}))},$$

where $m = \min\{\delta(F(t,\alpha,\boldsymbol{a})), k\}$. Therefore, one has

$$\gamma_{\delta(\boldsymbol{a})}(Q(t,\alpha,\boldsymbol{a})) = \gamma_{\delta(\boldsymbol{a})}(\alpha(F(t,\alpha,\boldsymbol{a}))) \in N_m^n / K_{\delta(\boldsymbol{a})}.$$

Note that the assumptions on F imply that

$$\min\{\delta(F(t,\alpha,\boldsymbol{a})), k\} = \min\{\delta(\boldsymbol{a}), k\}.$$

Hence $Q(t,\alpha,\boldsymbol{a}) \in X_k$.

It is seen straightaway that Q is a function over Δ_n. Since F is N_0^n-invariant, and since the functions $\gamma_{\delta(\boldsymbol{a})}$ and $e_{\delta(\boldsymbol{a})}$ are G-equivariant, it follows that Q is N_0^n-equivariant. What remains is to show that Q is continuous. Let us unleash the full power of Lemma 2.2.

Observe that $\varpi \circ Q = \mathrm{pr}$, so it is surely continuous. Let $0 \leq m \leq n$. Now $(\varpi \circ Q)^{-1}(\Delta_n \setminus \Delta_{m-1}) = I \times \mathbf{CS}(X_k) \times (\Delta_n \setminus \Delta_{m-1})$ and

$$\begin{array}{rccc} \gamma_m \circ Q\colon & I \times \mathbf{CS}(X_k) \times (\Delta_n \setminus \Delta_{m-1}) & \to & G/K_m \\ & (t,\alpha,\boldsymbol{a}) & \mapsto & \gamma_m(\alpha(F(t,\alpha,\boldsymbol{a}))) \end{array}.$$

Since Δ_n is a compact Hausdorff space, the evaluation $\mathrm{e}\colon \mathbf{CS}(X) \times \Delta_n \to X$ is continuous by Proposition 3.3. Now $\gamma_m \circ Q = \gamma_m \circ \mathrm{e} \circ (\mathrm{pr}, F)$, so $\gamma_m \circ Q$ is continuous. We now infer from Lemma 2.2 that Q is continuous.

Continuity of the adjoint $\hat{Q}$ follows by Proposition 3.4. □

5. Equivariant deformation retracts

The notion of equivariant deformation retract is central to the main results of this thesis. In this section we concentrate on the special case where a subgroup acts on a group by multiplication. We will explain how this relates to local cross-sections and local contraction properties of quotient spaces.

Let N be a topological group and let $M \subset N$ be a subgroup. Then N may be considered as an M-space, where the subgroup M acts from the left by multiplication $(m, g) \mapsto mg$. For this particular action it is best to denote the orbit space by $M \backslash N$ to distinguish it from the space N/M of left cosets of M in N. We consider the quotient map $\pi\colon N \to M \backslash N$.

Let us make a note on the neighbourhoods of Me in $M\backslash N$. If W is an open neighbourhood of Me in $M\backslash N$, then $V = \pi^{-1}(W)$ is an open M-invariant neighbourhood of M in N and it has the property that $M\backslash V = W$. Conversely, if V is an open M-invariant neighbourhood of M in N, then by openness of the quotient map the set $M\backslash V = W$ is an open neighbourhood of Me in $M\backslash N$ and it has the property that $V = \pi^{-1}(W)$.

5.1. Equivariant neighbourhood retracts and local cross-sections

We begin by recalling two basic notions.

1. We say that M is an *equivariant neighbourhood retract of* N if there exist an open M-invariant neighbourhood U of M in N and an M-map $r\colon U \to M$ (with the property that $r(g) = g$ for all $g \in M$).
2. We say that M has a *local cross-section in* N if the canonical map $\pi\colon N \to M\backslash N$ has a local cross-section $\sigma\colon M\backslash U \to N$ at Me (such that $\sigma(Me) = e$) for some open M-invariant neighbourhood U of M in N.

The conditions in parentheses are redundant in the following sense: if $r'\colon U \to M$ is an M-map, then r can be defined by setting $r(g) = r'(g)r'(e)^{-1}$ and if $\sigma'\colon M\backslash U \to N$ is a cross-section at Me, then σ can be defined by setting $\sigma(Mg) = \sigma'(Me)^{-1}\sigma'(Mg)$.

These above properties are in fact equivalent. Suppose that $r\colon U \to M$ is M-equivariant and satisfies $r(g) = g$ for all $g \in M$. The canonical map $\pi\colon U \to M\backslash U$ is open and hence a quotient map onto the open neighbourhood $M\backslash U$ of Me in $M\backslash N$. The map

$$\begin{aligned} s\colon \quad U &\to \quad N \\ g &\mapsto \quad r(g)^{-1}g \end{aligned}$$

is surely M-invariant and hence it factors through $M\backslash U$ inducing a continuous map $\sigma\colon M\backslash U \to N$, a local cross-section of π such that $\sigma(Me) = e$.

Conversely, if $\sigma\colon M\backslash U \to N$ is a local cross-section such that $\sigma(Me) = e$, then the map

$$\begin{aligned} r\colon \quad U &\to \quad M \\ g &\mapsto \quad g\sigma(Mg)^{-1} \end{aligned}$$

is M-equivariant and has the property that $r(g) = g$ for all $g \in M$.

Therefore, M is an equivariant neighbourhood retract of N if and only if M has a local cross-section in N.

5.2. Strong equivariant neighbourhood deformation retracts

We say that the subgroup M is a *strong equivariant neighbourhood deformation retract of* N if there exist an open M-invariant neighbourhood V of M in N and an M-map $D\colon I \times V \to N$ with the following properties:

1. $D(0, g) = g$ for all g;
2. $D(t, g) = g$ for all t and all $g \in M$; and

3. $D(1, g) \in M$ for all g.

Here too the second condition is redundant in the following sense. If

$$D'\colon I \times V \to N$$

is an M-map satisfying 1 and 3, then

$$D(t, g) = D'(t, g)D'(t, e)^{-1}$$

satisfies all three conditions. To make the exposition easier to follow, we will only work with equivariant deformations that satisfy 2.

The quotient space $M\backslash N$ is said to be *locally contractible* if for every open neighbourhood $M\backslash U$ of Me there exist an open neighbourhood $M\backslash W$ of Me and a deformation $D'\colon I \times M\backslash W \to M\backslash U$ such that

1. $D'(0, Mg) = Mg$ for all Mg; and
2. $D'(1, Mg) = Me$ for all Mg.

Theorem 5.1. *Let N be a topological group and let $M \subset N$ be a subgroup. Then M is a strong equivariant neighbourhood deformation retract of N if and only if*

1. *M has a local cross-section in N; and*
2. *$M\backslash N$ is locally contractible.*

Proof. Let us first show that the two conditions are necessary. Let $D\colon I \times V \to N$ be a strong equivariant deformation retraction onto M with V open in N. Since the map

$$\begin{array}{rccc} r\colon & V & \to & M \\ & g & \mapsto & D(1, g) \end{array}$$

is an equivariant retraction, there exists a local cross-section $\sigma\colon M\backslash V \to N$, so the first condition holds.

Since $\pi\colon N \to M\backslash N$ is an open map, so is the map $\pi\colon V \to M\backslash V$ and hence also

$$\mathrm{id} \times \pi\colon I \times V \to I \times M\backslash V.$$

Since D is an M-equivariant strong deformation retraction, it induces a continuous map $D'\colon I \times M\backslash V \to M\backslash N$ with the following properties:

1. $D'(0, Mg) = Mg$ for all Mg;
2. $D'(t, Me) = Me$ for all t; and
3. $D'(1, Mg) = Me$ for all Mg.

If $M\backslash U$ is an open neighbourhood of $Me \in M\backslash N$, then $D'^{-1}(M\backslash U)$ is an open neighbourhood of $I \times \{Me\} \subset I \times M\backslash V$. By Lemma 1.12, there exists an open neighbourhood $M\backslash W$ of Me such that $D'(I \times M\backslash W) \subset M\backslash U$. This shows that the second condition also holds, that is, $M\backslash N$ is locally contractible.

Now suppose that M and N satisfy the two given conditions. Since M has a local cross-section in N, there exist an open M-invariant neighbourhood U of M in N and a cross-section $\sigma\colon M\backslash U \to N$ of the map $\pi\colon N \to M\backslash N$. Moreover, one can always require that $\sigma(Me) = e$. Since $M\backslash N$ is locally contractible, U contains an open M-invariant neighbourhood V of M such that there exists a continuous map $D'\colon I \times M\backslash V \to M\backslash U$ with the following properties:

1. $D'(0, Mg) = Mg$ for all Mg; and
2. $D'(1, Mg) = Me$ for all Mg.

Now consider the continuous map

$$\begin{array}{rccc} D\colon & I \times V & \to & N \\ & (t, g) & \mapsto & g\sigma(Mg)^{-1}\sigma(D'(t, Mg))\sigma(D'(t, Me))^{-1} \end{array}.$$

It is certainly M-equivariant and it has the following properties:

1. $D(0, g) = g\sigma(Mg)^{-1}\sigma(D'(0, Mg))\sigma(D'(0, Me))^{-1}$
$= g\sigma(Mg)^{-1}\sigma(Mg)\sigma(Me)^{-1} = ge^{-1} = g$ holds for all g.
2. $D(t, g) = g\sigma(Mg)^{-1}\sigma(D'(t, Mg))\sigma(D'(t, Me))^{-1}$
$= g\sigma(Me)^{-1}\sigma(D'(t, Me))\sigma(D'(t, Me))^{-1} = ge^{-1} = g$ holds for all t and all $g \in M$.
3. $D(1, g) = g\sigma(Mg)^{-1}\sigma(D'(1, Mg))\sigma(D'(1, Me))^{-1}$
$= g\sigma(Mg)^{-1}\sigma(Me)\sigma(Me)^{-1} = g\sigma(Mg)^{-1} \in M$ holds for all g. □

Remark 5.2. Observe that it can be seen from the proof that M is a strong equivariant deformation retract of N if and only if M has a global cross-section in N and the quotient space $M\backslash N$ is contractible. Moreover, in such a situation the point Me is a strong deformation retract of $M\backslash N$.

6. The N_0^n-space $\mathbf{CS}(X_0)$ is a strong N_0^n-deformation retract of $\mathbf{CS}(X_n)$

For the remaining part of the thesis, G is a metrisable group, $K_0 \supset \cdots \supset K_n$ are compact subgroups and X denotes the equivariant simplex $\Delta_n(G; K_0, \ldots, K_n)$, which we consider as a left G-space. By X_k with $0 \leq k \leq n$ we denote the N_0^n-invariant subspace of X defined in Section 4.2. The space $\mathbf{CS}(X_k)$ is considered as a left N_0^n-space, see Section 4.4.

The aim of this section is to show that if G is metrisable and if N_{k-1}^n is a strong equivariant neighbourhood deformation retract of N_k^n for all $1 \leq k \leq n$, then $\mathbf{CS}(X_0)$ is a strong N_0^n-deformation retract of $\mathbf{CS}(X_n)$, see Corollary 6.10. Obviously, it is enough to show separately for each $1 \leq k \leq n$ that $\mathbf{CS}(X_{k-1})$ is a strong N_0^n-deformation retract of $\mathbf{CS}(X_k)$ if N_{k-1}^n is a strong equivariant neighbourhood deformation retract of N_k^n, which we do in Theorem 6.9.

The proof of Theorem 6.9 is far from simple. It has an intermediate state where $\mathbf{CS}(X_k)$ is deformed into $\mathbf{CS}(W_k)$, where W_k is a certain N_0^n-invariant subspace of X_k which contains X_{k-1}. This deformation is stationary on $\mathbf{CS}(X_{k-1})$.

6.1. The N_0^n-spaces W_k

The aim of this section is to construct for any $1 \leq k \leq n$ an N_0^n-invariant subspace W_k of X_k having the properties listed in Theorem 6.4. Notice that X_{k-1} is a strong N_0^n-deformation retract of W_k. However, W_k is *not* a *neighbourhood* of the *whole* X_{k-1}.

Proposition 6.1. *Let $\delta(\boldsymbol{a}) \geq k-1$. Then*

$$N_k^n \cap K_{\delta(\boldsymbol{a})} = N_{k-1}^n \cap K_{\delta(\boldsymbol{a})}.$$

Proof. Since $N_{k-1}^n \subset N_k^n$, it follows that $N_{k-1}^n \cap K_{\delta(\boldsymbol{a})} \subset N_k^n \cap K_{\delta(\boldsymbol{a})}$. Moreover, $K_{\delta(\boldsymbol{a})} \subset K_{k-1} \subset N_{k-1}$, so $N_k^n \cap K_{\delta(\boldsymbol{a})} \subset N_{k-1}^n$. Therefore, $N_k^n \cap K_{\delta(\boldsymbol{a})} \subset N_{k-1}^n \cap K_{\delta(\boldsymbol{a})}$. □

We recall from Section 4.2 that X^k is the subspace of $N_k^n \times \Delta_n$ consisting of those points that p maps to X_k and that $p\colon X^k \to X_k$ is a proper surjective map.

Proposition 6.2. *Let $(g, \boldsymbol{a}) \in X^k$ and suppose that $\delta(\boldsymbol{a}) = k-1$. Then $g \in N_{k-1}^n$.*

Proof. It follows from the definition of the set X^k that $g \in N_k^n$ and $gK_{k-1} \in N_{k-1}^n/K_{k-1}$. Therefore, there exists a $g' \in N_{k-1}^n$ such that $g'^{-1}g \in K_{k-1}$. Now $g'^{-1}g \in N_k^n$, so that by Proposition 6.1, $g'^{-1}g \in N_{k-1}^n$. Hence $g \in N_{k-1}^n$. □

We define for each $1 \leq k \leq n$ the continuous map

$$\begin{array}{rccc} w_{k-1}\colon & \Delta_n & \to & I \\ & \boldsymbol{a} & \mapsto & a_{k-1} + \cdots + a_n \end{array},$$

for which $w_{k-1}(\boldsymbol{a}) = 0$ if and only if $\delta(\boldsymbol{a}) < k-1$. This map will have an important role throughout Section 6.

Proposition 6.3. *Let $1 \leq k \leq n$. Let V be an N_{k-1}^n-invariant neighbourhood of N_{k-1}^n in N_k^n. Then there exist N_0^n-invariant sets U and W such that*

1. $X^{k-1} \subset W \subset X^k$;
2. $N_{k-1}^n \times (\Delta_n \setminus \Delta_{k-2}) \subset U \subset W|(\Delta_n \setminus \Delta_{k-2})$;
3. *U is open and p-saturated in X^k and W is closed in X^k;*
4. $W \subset V \times \Delta_n$;
5. *$W|\Delta_{k-1} \subset N_{k-1}^n \times \Delta_{k-1}$; and*
6. *if $(g, \boldsymbol{a}') \in W$, $\delta(\boldsymbol{a}) = \delta(\boldsymbol{a}')$ and $w_{k-1}(\boldsymbol{a}) = w_{k-1}(\boldsymbol{a}')$, then $(g, \boldsymbol{a}) \in W$.*

Proof. Since G is a metrisable group, it has a left G-invariant metric d, as was noted in the proof of Theorem 2.6. We confine the analysis to N_k^n. The distance of a point g from the closed set N_{k-1}^n gives a continuous N_{k-1}^n-invariant function. This has the property that $d(g, N_{k-1}^n) = 0$ if and only if $g \in N_{k-1}^n$.

Since V is a neighbourhood of N_{k-1}^n, it follows that there exists an $\epsilon > 0$ such that $d(g, N_{k-1}^n) \geq \epsilon$ if $g \notin V$. For otherwise one would find for each $\epsilon > 0$ points $g \notin V$ and $m \in N_{k-1}^n$ such that $d(g, m) < \epsilon$. Then by left invariance $d(m^{-1}g, e) < \epsilon$, which would imply that $e \in \overline{N_k^n \setminus V}$, contradicting the fact that V is a neighbourhood of $e \in N_{k-1}^n$. The function $\nu^k \colon N_k^n \to I$,

$$\nu^k(g) = \frac{\min\{\epsilon, d(g, N_{k-1}^n)\}}{\epsilon},$$

is a continuous N_{k-1}^n-invariant map and has the properties that

1. $\nu^k(g) = 0$ if and only if $g \in N_{k-1}^n$; and
2. $\nu^k(g) = 1$ for all $g \notin V$.

We now define

$$U = \{\ (g, \boldsymbol{a}) \in X^k \mid \nu^k(g) < w_{k-1}(\boldsymbol{a})/2\ \}$$

and

$$W = \{\ (g, \boldsymbol{a}) \in X^k \mid \nu^k(g) \leq w_{k-1}(\boldsymbol{a})/2\ \}.$$

Since ν^k is N_{k-1}^n-invariant and hence also N_0^n-invariant, both U and W are N_0^n-invariant sets.

Since $X^{k-1} \subset N_{k-1}^n \times \Delta_n$ and since $\nu^k(g) = 0$ for all $g \in N_{k-1}^n$, it follows that Property 1 holds. Similarly, since $w_{k-1}(\boldsymbol{a}) > 0$ if and only if $\boldsymbol{a} \in \Delta_n \setminus \Delta_{k-2}$ and since $\nu^k(g) = 0$ for all $g \in N_{k-1}^n$, it follows that Property 2 holds.

Since w_{k-1} and ν^k are continuous maps, it follows that U is open in X^k and W is closed in X^k. That U is p-saturated in X^k is somewhat more difficult to show. Let $(g, \boldsymbol{a}) \in U$ and let $(g', \boldsymbol{a}') \in X^k$ be such that $\langle g, \boldsymbol{a} \rangle = \langle g', \boldsymbol{a}' \rangle$. It immediately follows that $\boldsymbol{a}' = \boldsymbol{a}$ and that ${g'}^{-1}g \in K_{\delta(\boldsymbol{a})}$. It suffices to show that $\nu^k(g) = \nu^k(g')$.

It follows from the definition of the set U that $\delta(\boldsymbol{a}) \geq k-1$. If $\delta(\boldsymbol{a}) = k-1$, then it follows from Proposition 6.2 that both g and g' belong to N_{k-1}^n. Hence $\nu^k(g) = 0 = \nu^k(g')$.

The complementary situation $\delta(\boldsymbol{a}) \geq k$ requires a different type of argument, in which the assumption ${g'}^{-1}g \in K_{\delta(\boldsymbol{a})}$ is required. Now $g = g'({g'}^{-1}g)$, and since ${g'}^{-1}g \in K_{\delta(\boldsymbol{a})}$ and $g' \in N_{\delta(\boldsymbol{a})}$, by normality there exists an $l \in K_{\delta(\boldsymbol{a})}$ such that $lg' = g$. Since $l = g{g'}^{-1}$, it follows that $l \in N_k^n$. Now recall that by Proposition 6.1

$$N_k^n \cap K_{\delta(\boldsymbol{a})} = N_{k-1}^n \cap K_{\delta(\boldsymbol{a})},$$

which implies that $l \in N_{k-1}^n$. Since ν^k is N_{k-1}^n-invariant, it follows that $\nu^k(g) = \nu^k(lg') = \nu^k(g')$. This completes the proof that U is p-saturated in X^k and hence Property 3 holds.

Since $\nu^k(g) \leq w_{k-1}(\boldsymbol{a})/2 < 1$ for $(g, \boldsymbol{a}) \in W$, it follows from the properties of ν^k that $g \in V$. Hence Property 4 holds. If $(g, \boldsymbol{a}) \in W|\Delta_{k-1}$, then either $\delta(\boldsymbol{a}) = k-1$ so that $g \in N_{k-1}^n$ by Proposition 6.2 or $\delta(\boldsymbol{a}) < k-1$ so that $\nu^k(g) \leq w_{k-1}(\boldsymbol{a})/2 = 0$, which also leads to the conclusion that $g \in N_{k-1}^n$. Hence Property 5 holds.

To check that Property 6 holds, first note that since $(g, \boldsymbol{a}') \in X^k$, the condition $\delta(\boldsymbol{a}) = \delta(\boldsymbol{a}')$ implies that $(g, \boldsymbol{a}) \in X^k$ also, and then observe that the condition $w_{k-1}(\boldsymbol{a}) = w_{k-1}(\boldsymbol{a}')$ further implies that $(g, \boldsymbol{a}) \in W$. □

We proceed to define the N_0^n-spaces W_k.

Theorem 6.4. *Let $1 \leq k \leq n$. Suppose that N^n_{k-1} is a strong equivariant neighbourhood deformation retract of N^n_k. Then there exists an N^n_0-invariant set W_k such that*

1. $X_{k-1} \subset W_k \subset X_k$;
2. *if* $\alpha \in \mathbf{CS}(X_k)$, $\boldsymbol{a}, \boldsymbol{a}' \in \Delta_n$, $\alpha(\boldsymbol{a}') \in W_k$, $\delta(\boldsymbol{a}) = \delta(\boldsymbol{a}')$ *and* $w_{k-1}(\boldsymbol{a}) = w_{k-1}(\boldsymbol{a}')$, *then* $e_{\delta(\boldsymbol{a})}(\gamma_{\delta(\boldsymbol{a})}(\alpha(\boldsymbol{a}')), \boldsymbol{a}) \in W_k$;
3. $X_{k-1}|(\Delta_n \setminus \Delta_{k-2}) \subset \operatorname{int} W_k|(\Delta_n \setminus \Delta_{k-2})$ *in* $X_k|(\Delta_n \setminus \Delta_{k-2})$; *and*
4. *there exists an* N^n_0*-map* $H_k \colon I \times W_k \to X_k$ *over* Δ_n *such that*
   (i) $H_k(0, \langle g, \boldsymbol{a}\rangle) = \langle g, \boldsymbol{a}\rangle$ *for all* $\langle g, \boldsymbol{a}\rangle$;
   (ii) $H_k(t, \langle g, \boldsymbol{a}\rangle) = \langle g, \boldsymbol{a}\rangle$ *for all* t *and all* $\langle g, \boldsymbol{a}\rangle \in X_{k-1}$; *and*
   (iii) $H_k(1, \langle g, \boldsymbol{a}\rangle) \in X_{k-1}$ *for all* $\langle g, \boldsymbol{a}\rangle$.

Proof. We recall from Section 5 that since N^n_{k-1} is a strong equivariant neighbourhood deformation retract of N^n_k, there exists an N^n_{k-1}-invariant neighbourhood V^k of N^n_{k-1} in N^n_k and an N^n_{k-1}-equivariant map $D \colon I \times V^k \to N^n_k$ with the following properties:

1. $D(0, g) = g$ for all g;
2. $D(t, g) = g$ for all t and all $g \in N^n_{k-1}$; and
3. $D(1, g) \in N^n_{k-1}$ for all g.

Let $U = U^k$ and $W = W^k$ be the sets as in Proposition 6.3 corresponding to the set $V = V^k$. We define $W_k = p(W^k)$. Since W^k is an N^n_0-space, so is W_k. Since W^k is a closed subset of X^k and since $p \colon X^k \to X_k$ is a proper map, W_k is closed in X_k. Moreover, $p \colon W^k \to W_k$ is a proper map by Proposition 1.9.1 since W^k is closed in X^k. Since $X^{k-1} \subset W^k \subset X^k$ by Proposition 6.3.1, it is evident that W_k satisfies Property 1.

To show that W_k satisfies Property 2, note that $\langle g, \boldsymbol{a}'\rangle = \alpha(\boldsymbol{a}') \in W_k$ for some $(g, \boldsymbol{a}') \in W^k$. Since $\delta(\boldsymbol{a}) = \delta(\boldsymbol{a}')$ and $w_{k-1}(\boldsymbol{a}) = w_{k-1}(\boldsymbol{a}')$, it follows from Proposition 6.3.6 that $(g, \boldsymbol{a}) \in W^k$ and hence $\langle g, \boldsymbol{a}\rangle \in W_k$. But since $\delta(\boldsymbol{a}) = \delta(\boldsymbol{a}')$, it follows by Proposition 2.3 that $e_{\delta(\boldsymbol{a})}(\gamma_{\delta(\boldsymbol{a})}(\alpha(\boldsymbol{a}')), \boldsymbol{a}) = e_{\delta(\boldsymbol{a})}(gK_{\delta(\boldsymbol{a})}, \boldsymbol{a}) = \langle g, \boldsymbol{a}\rangle \in W_k$, as required.

Observe that

$$X^{k-1}|(\Delta_n \setminus \Delta_{k-2}) \subset N^n_{k-1} \times (\Delta_n \setminus \Delta_{k-2}) \subset U^k \subset W^k|(\Delta_n \setminus \Delta_{k-2}).$$

Since U^k is an open p-saturated set in X^k, the set $U_k = p(U^k)$ is open in X_k. It follows that Property 3 holds. Note that in fact $U_k \subset X_k|(\Delta_n \setminus \Delta_{k-2})$.

We define an N^n_0-equivariant map over Δ_n as follows:

$$\begin{array}{cccc} H^k \colon & I \times W^k & \to & X^k \\ & (t, (g, \boldsymbol{a})) & \mapsto & (D(t, g), \boldsymbol{a}) \end{array}.$$

Let us first show that H^k is well-defined. Since $W^k \subset V^k \times \Delta_n$ by Proposition 6.3.4, $D(t, g)$ is a well-defined element of N^n_k. We also need to check that $(D(t, g), \boldsymbol{a}) \in X^k$.

If $\delta(\boldsymbol{a}) < k$, then $g \in N^n_{k-1}$ by Proposition 6.3.5. Hence $D(t,g) = g$ so that $(D(t,g),\boldsymbol{a}) \in X^k$. If $\delta(\boldsymbol{a}) \geq k$, it suffices to note that $D(t,g) \in N^n_k$.

Observe that H^k has the following properties:

1. $H^k(0,(g,\boldsymbol{a})) = (D(0,g),\boldsymbol{a}) = (g,\boldsymbol{a})$ for all $(g,\boldsymbol{a})$;
2. $H^k(t,(g,\boldsymbol{a})) = (D(t,g),\boldsymbol{a}) = (g,\boldsymbol{a})$ for all t and all $(g,\boldsymbol{a}) \in X^{k-1}$ since $g \in N^n_{k-1}$; and
3. $H^k(1,(g,\boldsymbol{a})) \in X^{k-1}$ for all $(g,\boldsymbol{a})$. For if $(g,\boldsymbol{a}) \in W^k$ and $\delta(\boldsymbol{a}) \leq k-1$, then $g \in N^n_{k-1}$ and hence $D(1,g) = g$, and if $\delta(\boldsymbol{a}) > k-1$, then $D(1,g) \in N^n_{k-1}$.

We next show that H^k induces an N^n_0-equivariant map over Δ_n as follows:

$$\begin{array}{cccc} H_k\colon & I \times W_k & \longrightarrow & X_k \\ & (t,\langle g,\boldsymbol{a}\rangle) & \longmapsto & \langle D(t,g),\boldsymbol{a}\rangle \end{array}.$$

Since $p\colon W^k \longrightarrow W_k$ is a proper map, it suffices to show that $p \circ H^k$ factors through $\mathrm{id} \times p$, which is a closed map and hence a quotient map.

Let $\langle g,\boldsymbol{a}\rangle = \langle g',\boldsymbol{a}\rangle$. The situation $\delta(\boldsymbol{a}) < k$ is rather evident. For now $g, g' \in N^n_{k-1}$ so that

$$\langle D(t,g),\boldsymbol{a}\rangle = \langle g,\boldsymbol{a}\rangle = \langle g',\boldsymbol{a}\rangle = \langle D(t,g'),\boldsymbol{a}\rangle.$$

The complementary situation $\delta(\boldsymbol{a}) \geq k$ requires a different type of argument. Now $g, g' \in N^n_k$ and $g'^{-1}g \in K_{\delta(\boldsymbol{a})}$. Then $g = g'(g'^{-1}g)$, and since $g'^{-1}g \in K_{\delta(\boldsymbol{a})}$ and $g' \in N_{\delta(\boldsymbol{a})}$, there exists an $l \in K_{\delta(\boldsymbol{a})}$ such that $lg' = g$. Since $l = gg'^{-1}$, it follows that $l \in N^n_k$. Now recall that by Proposition 6.1

$$N^n_k \cap K_{\delta(\boldsymbol{a})} = N^n_{k-1} \cap K_{\delta(\boldsymbol{a})},$$

which implies that $l \in N^n_{k-1}$. Since $l \in K_{\delta(\boldsymbol{a})}$ and since $D(t,g') \in N_{\delta(\boldsymbol{a})}$, there exists an $l' \in K_{\delta(\boldsymbol{a})}$ such that $lD(t,g') = D(t,g')l'$. Therefore,

$$D(t,g) = D(t,lg') = lD(t,g') = D(t,g')l',$$

which shows that

$$\langle D(t,g),\boldsymbol{a}\rangle = \langle D(t,g')l',\boldsymbol{a}\rangle = \langle D(t,g'),\boldsymbol{a}\rangle.$$

The three stated properties of H_k are direct consequences of the corresponding properties of H^k. □

6.2. The N^n_0-deformation of $\mathbf{CS}(X_k)$ into $\mathbf{CS}(W_k)$ stationary on $\mathbf{CS}(X_{k-1})$

In Section 6.1 we defined an N^n_0-invariant subspace W_k of X_k containing X_{k-1} as its strong N^n_0-deformation retract. In order to apply this deformation, one has to first deform $\mathbf{CS}(X_k)$ into $\mathbf{CS}(W_k)$.

Our construction of such a deformation makes use of the fact that if α is a cross-section of X_k, then $\alpha(\Delta_{k-1}) \subset X_{k-1}$, as follows from Proposition 4.2. The idea is then to use this information to find for each cross-section α a neighbourhood of $\Delta_{k-1} \setminus \Delta_{k-2}$ where the values of α belong to W_k. Our purpose being to apply Proposition 4.4, we will construct a dimension preserving parametrised deformation that compresses the simplex Δ_n to that neighbourhood. This, however, requires that the neighbourhoods have been chosen in a continuous way in respect of the parameter α. Here Proposition 6.7 plays a key role.

We start by defining a parametrised deformation of Δ_n. It squeezes the simplex Δ_n as close to Δ_{k-1} as is needed while holding Δ_{k-1} pointwise fixed.

Lemma 6.5. *Let $1 \le k \le n$. There exists a map*

$$C_k \colon I \times \Delta_n \times I \to \Delta_n$$

with the following properties:

1. $C_k(0, \boldsymbol{a}, s) = \boldsymbol{a}$ *for all* $\boldsymbol{a}$ *and* s;
2. $C_k(t, \boldsymbol{a}, s) = \boldsymbol{a}$ *for all* t, *all* $\boldsymbol{a} \in \Delta_{k-1}$ *and all* s;
3. $C_k(t, \boldsymbol{a}, 1) = \boldsymbol{a}$ *for all* t *and* $\boldsymbol{a}$;
4. $C_k(1, \boldsymbol{a}, 0) \in \Delta_{k-1}$ *for all* $\boldsymbol{a}$;
5. $w_{k-1}(C_k(t, \boldsymbol{a}, s)) = w_{k-1}(\boldsymbol{a})$ *for all* t, *all* $\boldsymbol{a}$ *and all* s; *and*
6. $\delta(C_k(t, \boldsymbol{a}, s)) = \delta(\boldsymbol{a})$ *for all* t, *all* $\boldsymbol{a}$ *and all* $s > 0$.

Proof. We define C_k by giving its component functions

$$C_k(t, \boldsymbol{a}, s)_l = \begin{cases} a_l, & \text{when } 0 \le l \le k-2; \\ a_l + t(1-s)(a_k + \cdots + a_n), & \text{when } l = k-1; \text{ and} \\ a_l - t(1-s)a_l, & \text{when } k \le l \le n. \end{cases}$$

This gives a well-defined continuous map, which has Properties 1 and 3.

If $\boldsymbol{a} \in \Delta_{k-1}$, then $a_l = 0$ when $k \le l \le n$. Hence $C_k(t, \boldsymbol{a}, s)_l = a_l$ when $0 \le l \le n$, which implies that $C_k(t, \boldsymbol{a}, s) = \boldsymbol{a}$ as required in Property 2.

When $k \le l \le n$, then

$$C_k(1, \boldsymbol{a}, 0)_l = a_l - 1(1-0)a_l = 0,$$

which implies that $C_k(1, \boldsymbol{a}, 0) \in \Delta_{k-1}$ as required in Property 4.

Property 5 is easily checked by summing the components of $C_k(t, \boldsymbol{a}, s)$ for $k-1 \le l \le n$.

Property 6 is simple, but has many steps. Recall that

$$\delta(\boldsymbol{a}) = \max\{\, 0 \le l \le n \mid a_l > 0 \,\}.$$

Let $\delta(\boldsymbol{a}) = l$. Then $a_l > 0$ and $a_{l'} = 0$ for all $l' > l$.

If $k \le l \le n$, then $C_k(t, \boldsymbol{a}, s)_l = (1 - t(1-s))a_l > 0$ since $1 - t(1-s) > 0$, and $C_k(t, \boldsymbol{a}, s)_{l'} = (1 - t(1-s))a_{l'} = 0$ for all $l' > l$. Hence $\delta(C_k(t, \boldsymbol{a}, s)) = l$.

If $l = k-1$, then $C_k(t, \boldsymbol{a}, s)_l = a_l + t(1-s)(a_k + \cdots + a_n) = a_l > 0$ and $C_k(t, \boldsymbol{a}, s)_{l'} = (1 - t(1-s))a_{l'} = 0$ for all $l' > l$. Hence $\delta(C_k(t, \boldsymbol{a}, s)) = l$.

If $0 \le l \le k-2$, then $C_k(t, \boldsymbol{a}, s)_l = a_l > 0$ and $C_k(t, \boldsymbol{a}, s)_{l'} = a_{l'} = 0$ for all $l < l' \le k-2$, $C_k(t, \boldsymbol{a}, s)_{k-1} = a_{k-1} + t(1-s)(a_k + \cdots + a_n) = 0$ and $C_k(t, \boldsymbol{a}, s)_{l'} = (1 - t(1-s))a_{l'} = 0$ for all $k \le l' \le n$. Hence in this case too $\delta(C_k(t, \boldsymbol{a}, s)) = l$. □

Recall that a function $f\colon X \to \mathbb{R}$ is said to be *upper semicontinuous* if $f^{-1}(]-\infty, a[)$ is open for all $a \in \mathbb{R}$ and *lower semicontinuous* if $f^{-1}(]a, \infty[)$ is open for all $a \in \mathbb{R}$. Characteristic functions of open sets are lower semicontinuous functions.

Lemma 6.6. *Let X be a metrisable space and let $A \subset X$ be a closed subspace. Let $\tau\colon X \times I \to \mathbb{R}$ be a lower semicontinuous function such that*

1. *$\tau(x, 0) = 1$ for all x; and*
2. *$\tau(x, s) = 1$ for all $x \in A$ and all s.*

Then there exists a continuous map $\sigma\colon X \to]0, 1]$ such that

1. *$\tau(x, s) > 0$ for all $s \leq \sigma(x)$; and*
2. *$\sigma(x) = 1$ for all $x \in A$.*

Proof. We define for each $r \in]0, 1]$ a set

$$U(r) = \{\ x \in X \mid \tau(x, s) > 0 \text{ for all } s \leq r\ \}.$$

Let us first show that the sets $U(r)$ are open. For this, let us denote $W = \tau^{-1}(]0, \infty[)$, which is open in $X \times I$ since τ is lower semicontinuous. It is obvious that a point x belongs to $U(r)$ if and only if $\{x\} \times [0, r] \subset W$. Since $[0, r]$ is compact, it follows from Lemma 1.12 that x has a neighbourhood U such that $U \times [0, r] \subset W$. It follows that $U(r)$ is open since $U \subset U(r)$.

We next show that the sets $U(r)$ form a cover of X. Let $x \in X$. Since $\tau(x, 0) = 1$, it follows that $(x, 0) \in W$. Since W is open, it contains a neighbourhood $U \times V$ of $(x, 0)$. If $r > 0$ is chosen so that $[0, r] \subset V$, then $x \in U(r)$.

We now define

$$\begin{array}{rccc} h\colon & X & \to & \mathbb{R} \\ & x & \mapsto & \sup\{\ r \mid x \in U(r)\ \} \end{array},$$

and we claim that h is lower semicontinuous. Let $a \in \mathbb{R}$ and let $x \in h^{-1}(]a, \infty[)$. Then $a < h(x)$, and hence there exists an r such that $a < r \leq h(x)$ and $x \in U(r)$. Now $U(r)$ is a neighbourhood of x. Moreover, if $x' \in U(r)$, then $a < r \leq h(x')$ so that $U(r) \subset h^{-1}(]a, \infty[)$. Hence $h^{-1}(]a, \infty[)$ is open and so h is a lower semicontinuous function on X.

Let $g\colon X \to \mathbb{R}$ denote the constant function $x \mapsto 0$, which is an upper semicontinuous function on X. Since h is strictly positive, $g < h$. Since X is a metrisable space, it is in particular countably paracompact and normal. By a well-known theorem, see [Dow51], Theorem 4β, there exists a continuous function $f\colon X \to \mathbb{R}$ such that $g < f < h$.

Notice next that if $x \in A$, then $x \in U(1)$ since $\tau(x, s) = 1 > 0$ for all $s \leq 1$. It follows that $U(1)$ is an open neighbourhood of the closed set A. Hence, by normality, see [Kel55], Lemma 4.4, there exists a continuous function $f'\colon X \to [0, 1]$ such that

1. $f'(x) = 1$ for all $x \in A$; and
2. $f'(x) = 0$ for all $x \in X \setminus U(1)$.

Let us finally define a continuous map as follows:

$$\begin{array}{rcl} \sigma\colon\ X & \to &]0,1] \\ x & \mapsto & \max\{f(x), f'(x)\} \end{array}.$$

Let x and s be such that $s \leq \sigma(x)$. Observe that $f'(x) > f(x)$ is possible only if $x \in U(1)$. Since $s \leq 1$, it follows from the definition of the set $U(1)$ that $\tau(x,s) > 0$. If on the other hand $f'(x) \leq f(x)$, then $s \leq \sigma(x) = f(x) < h(x)$. Hence there exists an $r > s$ such that $x \in U(r)$. Since $s \leq r$, it follows from the definition of the set $U(r)$ that $\tau(x,s) > 0$.

Since $\sigma(x) = f'(x) = 1$ for all $x \in A$, the proof is complete. □

Proposition 6.7. *Let X be a G-space having a metrisable orbit space and let Y be a G-space. Let $A \subset X$ be a closed G-invariant subspace and let $U \subset Y$ be an open G-invariant subspace. Let $\upsilon\colon X \times I \to Y$ be a G-map such that*

1. *$\upsilon(x,0) \in U$ for all x; and*
2. *$\upsilon(x,s) \in U$ for all $x \in A$ and all s.*

Then there exists a continuous G-invariant map $\sigma\colon X \to]0,1]$ such that

1. *$\upsilon(x,s) \in U$ for all $s \leq \sigma(x)$; and*
2. *$\sigma(x) = 1$ for all $x \in A$.*

Proof. Since A and U are G-invariant sets, the corresponding sets A/G and U/G are found on the orbit space level and they are closed and open, respectively. Let $\chi\colon Y/G \to \mathbb{R}$ be the characteristic function of the open set U/G, which means that

$$\chi(Gy) = \begin{cases} 1, & \text{when } Gy \in U/G\text{; and} \\ 0, & \text{when } Gy \in Y/G \setminus U/G. \end{cases}$$

Since υ is a G-map, it induces a continuous map $\upsilon'\colon X/G \times I \to Y/G$. Then

$$\tau = \chi \circ \upsilon'\colon X/G \times I \to \mathbb{R}$$

is a lower semicontinuous function such that

1. $\tau(Gx,0) = 1$ for all Gx; and
2. $\tau(Gx,s) = 1$ for all $Gx \in A/G$ and all s.

According to Lemma 6.6 there exists a continuous map $\tilde{\sigma}\colon X/G \to]0,1]$ such that

1. $\tau(Gx,s) > 0$ for all $s \leq \tilde{\sigma}(Gx)$; and
2. $\tilde{\sigma}(Gx) = 1$ for all $Gx \in A/G$.

By setting $\sigma(x) = \tilde{\sigma}(Gx)$ we obtain a continuous G-invariant map $\sigma\colon X \to]0,1]$. Let $s \leq \sigma(x)$. Observe that τ takes only two values, zero and one. Since $\tau(Gx,s) > 0$, it follows that $\chi(\upsilon'(Gx,s)) = \tau(Gx,s) = 1$ and hence $\upsilon'(Gx,s) \in U/G$. Therefore, $\upsilon(x,s) \in U$. Hence σ has the desired properties. □

Proposition 6.8. *Let $1 \leq k \leq n$. Let W_k be an N_0^n-invariant set such that*

1. *$X_{k-1} \subset W_k \subset X_k$;*

2. *if* $\alpha \in \mathbf{CS}(X_k)$, $\boldsymbol{a}, \boldsymbol{a}' \in \Delta_n$, $\alpha(\boldsymbol{a}') \in W_k$, $\delta(\boldsymbol{a}) = \delta(\boldsymbol{a}')$ *and* $w_{k-1}(\boldsymbol{a}) = w_{k-1}(\boldsymbol{a}')$, *then* $e_{\delta(\boldsymbol{a})}(\gamma_{\delta(\boldsymbol{a})}(\alpha(\boldsymbol{a}')), \boldsymbol{a}) \in W_k$; *and*
3. $X_{k-1}|(\Delta_n \setminus \Delta_{k-2}) \subset \operatorname{int} W_k|(\Delta_n \setminus \Delta_{k-2})$ *in* $X_k|(\Delta_n \setminus \Delta_{k-2})$.

Then there exists a continuous N_0^n*-equivariant map* $R_k\colon I \times \mathbf{CS}(X_k) \times \Delta_n \to X_k$ *over* Δ_n *such that*

1. $R_k(0, \alpha, \boldsymbol{a}) = \alpha(\boldsymbol{a})$ *for all* α *and* $\boldsymbol{a}$;
2. $R_k(t, \alpha, \boldsymbol{a}) = \alpha(\boldsymbol{a})$ *for all* t, *all* $\alpha \in \mathbf{CS}(X_{k-1})$ *and all* $\boldsymbol{a}$; *and*
3. $R_k(1, \alpha, \boldsymbol{a}) \in W_k$ *for all* α *and* $\boldsymbol{a}$.

Proof. Let $C_k\colon I \times \Delta_n \times I \to \Delta_n$ be a map which has the six properties that are listed in Lemma 6.5. Consider the N_0^n-map

$$\begin{array}{rccc} \upsilon\colon & \mathbf{CS}(X_k) \times (\Delta_n \setminus \Delta_{k-2}) \times I & \to & X_k|(\Delta_n \setminus \Delta_{k-2}) \\ & (\alpha, \boldsymbol{a}, s) & \mapsto & \alpha(C_k(1, \boldsymbol{a}, s)) \end{array}.$$

The map υ is well-defined since according to Property 5 of C_k it holds that

$$w_{k-1}(C_k(1, \boldsymbol{a}, s)) = w_{k-1}(\boldsymbol{a}) > 0$$

for all $\boldsymbol{a} \in \Delta_n \setminus \Delta_{k-2}$ and all s, and hence $\alpha(C_k(1, \boldsymbol{a}, s)) \in X_k|(\Delta_n \setminus \Delta_{k-2})$ for all $\boldsymbol{a} \in \Delta_n \setminus \Delta_{k-2}$ and all s. The map υ is continuous since C_k is continuous and since evaluation is continuous in $\mathbf{CS}(X_k) \times (\Delta_n \setminus \Delta_{k-2})$ by Proposition 3.3.

We recall from Section 4.4 that $\mathbf{CS}(X_k)$ is an N_0^n-invariantly metrisable proper N_0^n-space. One may shown as in Example 1.15 that $\mathbf{CS}(X_k) \times (\Delta_n \setminus \Delta_{k-2})$ is a proper N_0^n-space and obviously it is N_0^n-invariantly metrisable. Hence the N_0^n-space $\mathbf{CS}(X_k) \times (\Delta_n \setminus \Delta_{k-2})$ has a metrisable orbit space, see Remark 2.12.

Notice that

1. $\upsilon(\alpha, \boldsymbol{a}, s) \in X_{k-1}|(\Delta_n \setminus \Delta_{k-2})$ for all $\alpha \in \mathbf{CS}(X_{k-1})$, all $\boldsymbol{a}$ and all s; and
2. $\upsilon(\alpha, \boldsymbol{a}, 0) \in X_{k-1}|(\Delta_n \setminus \Delta_{k-2})$ for all α and all $\boldsymbol{a}$ by Proposition 4.2 since $C_k(1, \boldsymbol{a}, 0) \in \Delta_{k-1}$ according to Property 4 of C_k.

We now apply Proposition 6.7 to sets $A = \mathbf{CS}(X_{k-1}) \times (\Delta_n \setminus \Delta_{k-2})$ and $U = \operatorname{int} W_k|(\Delta_n \setminus \Delta_{k-2})$. That A is closed in $\mathbf{CS}(X_k) \times (\Delta_n \setminus \Delta_{k-2})$ follow from Theorem 4.3 and that υ maps $A \times I$ into U follows from Property 1 above and from Property 3 of W_k.

The resulting continuous N_0^n-invariant map

$$\sigma\colon\ \mathbf{CS}(X_k) \times (\Delta_n \setminus \Delta_{k-2}) \to]0, 1]$$

satisfies, in particular,

1. $\upsilon(\alpha, \boldsymbol{a}, \sigma(\alpha, \boldsymbol{a})) \in W_k$ for all α and all $\boldsymbol{a}$; and
2. $\sigma(\alpha, \boldsymbol{a}) = 1$ for all $\alpha \in \mathbf{CS}(X_{k-1})$ and all $\boldsymbol{a}$.

Since σ is N_0^n-invariant, the rule

$$F(t, \alpha, \boldsymbol{a}) = \begin{cases} C_k(t, \boldsymbol{a}, \sigma(\alpha, \boldsymbol{a})), & \text{when } \delta(\boldsymbol{a}) \geq k-1; \text{ and} \\ \boldsymbol{a}, & \text{when } \delta(\boldsymbol{a}) < k-1 \end{cases}$$

defines an N_0^n-invariant function $F\colon I \times \mathbf{CS}(X_k) \times \Delta_n \to \Delta_n$. Although F is clearly continuous in the open set where $\delta(\boldsymbol{a}) \geq k-1$, continuity of F at points where $\delta(\boldsymbol{a}) < k-1$ is not that evident. So let $\delta(\boldsymbol{a}) < k-1$ and let U be a neighbourhood of $F(t,\alpha,\boldsymbol{a}) = \boldsymbol{a}$. According to Property 2 of C_k, $C_k(t,\boldsymbol{a},s) = \boldsymbol{a}$ for all t, all $\boldsymbol{a} \in \Delta_{k-2}$ and all s, so it follows that

$$I \times \{\boldsymbol{a}\} \times I \subset C_k{}^{-1}(U).$$

By Lemma 1.12, $\boldsymbol{a}$ has a neighbourhood V in Δ_n such that

$$I \times V \times I \subset C_k{}^{-1}(U).$$

Intersecting V by U we may assume that $V \subset U$.

Now for all t', all α' and all $\boldsymbol{a}' \in V$ one either has that $\delta(\boldsymbol{a}') < k-1$ so that $F(t',\alpha',\boldsymbol{a}') = \boldsymbol{a}' \in U$, or $\delta(\boldsymbol{a}') \geq k-1$ so that $F(t',\alpha',\boldsymbol{a}') = C_k(t',\boldsymbol{a}',\sigma(\alpha',\boldsymbol{a}')) \in U$. This shows that F is continuous.

We next list five important properties of F. Note that they are trivial for all $\boldsymbol{a} \in \Delta_{k-2}$ so we can assume that $\boldsymbol{a} \in \Delta_n \setminus \Delta_{k-2}$.

1. Since $\sigma(\alpha,\boldsymbol{a}) > 0$, it follows from Property 6 of C_k that
$$\delta(F(t,\alpha,\boldsymbol{a})) = \delta(C_k(t,\boldsymbol{a},\sigma(\alpha,\boldsymbol{a}))) = \delta(\boldsymbol{a}).$$
2. It follows from Property 1 of C_k that
$$F(0,\alpha,\boldsymbol{a}) = C_k(0,\boldsymbol{a},\sigma(\alpha,\boldsymbol{a})) = \boldsymbol{a}.$$
3. If $\alpha \in \mathbf{CS}(X_{k-1})$, it follows from Property 3 of C_k that
$$F(t,\alpha,\boldsymbol{a}) = C_k(t,\boldsymbol{a},\sigma(\alpha,\boldsymbol{a})) = \boldsymbol{a},$$
since $\sigma(\alpha,\boldsymbol{a}) = 1$.
4. It follows from Property 1 of v that
$$\alpha(F(1,\alpha,\boldsymbol{a})) = \alpha(C_k(1,\boldsymbol{a},\sigma(\alpha,\boldsymbol{a}))) \in W_k.$$
5. It follows from Property 5 of C_k that
$$w_{k-1}(F(t,\alpha,\boldsymbol{a})) = w_{k-1}(C_k(t,\boldsymbol{a},\sigma(\alpha,\boldsymbol{a}))) = w_{k-1}(\boldsymbol{a}).$$

Proposition 4.4 then shows that

$$\begin{array}{rccc} R_k\colon & I \times \mathbf{CS}(X_k) \times \Delta_n & \to & X_k \\ & (t,\alpha,\boldsymbol{a}) & \mapsto & e_{\delta(\boldsymbol{a})}(\gamma_{\delta(\boldsymbol{a})}(\alpha(F(t,\alpha,\boldsymbol{a}))),\boldsymbol{a}) \end{array}$$

is a well-defined N_0^n-equivariant map over Δ_n. Let us then check that R_k has the required properties.

1. According to Property 2 of F,
$$R_k(0,\alpha,\boldsymbol{a}) = e_{\delta(\boldsymbol{a})}(\gamma_{\delta(\boldsymbol{a})}(\alpha(F(0,\alpha,\boldsymbol{a}))),\boldsymbol{a}) = e_{\delta(\boldsymbol{a})}(\gamma_{\delta(\boldsymbol{a})}(\alpha(\boldsymbol{a})),\boldsymbol{a}) = \alpha(\boldsymbol{a})$$
for all α and $\boldsymbol{a}$;

2. According to Property 3 of F,
$R_k(1,\alpha,\boldsymbol{a}) = e_{\delta(\boldsymbol{a})}(\gamma_{\delta(\boldsymbol{a})}(\alpha(F(1,\alpha,\boldsymbol{a}))),\boldsymbol{a}) = e_{\delta(\boldsymbol{a})}(\gamma_{\delta(\boldsymbol{a})}(\alpha(\boldsymbol{a})),\boldsymbol{a}) = \alpha(\boldsymbol{a})$
for all t, all $\alpha \in \mathbf{CS}(X_{k-1})$ and all $\boldsymbol{a}$; and
3. According to Property 4 of F, $\alpha(F(1,\alpha,\boldsymbol{a})) \in W_k$. According to Properties 1 and 5 of F, $\delta(\boldsymbol{a}) = \delta(F(1,\alpha,\boldsymbol{a}))$ and $w_{k-1}(\boldsymbol{a}) = w_{k-1}(F(1,\alpha,\boldsymbol{a}))$, so it follows from Property 2 of W_k that $R_k(1,\alpha,\boldsymbol{a}) = e_{\delta(\boldsymbol{a})}(\gamma_{\delta(\boldsymbol{a})}(\alpha(F(1,\alpha,\boldsymbol{a}))),\boldsymbol{a}) \in W_k$ for all α and $\boldsymbol{a}$. □

6.3. The combined inductive deformation

We are now ready to present the induction step, which is based on Theorem 6.4 and Proposition 6.8.

Theorem 6.9. *Let $1 \le k \le n$. Suppose that N^n_{k-1} is a strong equivariant neighbourhood deformation retract of N^n_k. Then $\mathbf{CS}(X_{k-1})$ is a strong N^n_0-deformation retract of $\mathbf{CS}(X_k)$.*

Proof. Since G is a metrisable group and since N^n_{k-1} is a strong equivariant neighbourhood deformation retract of N^n_k, it is shown in Theorem 6.4 that there exists an N^n_0-invariant set W_k such that

1. $X_{k-1} \subset W_k \subset X_k$;
2. if $\alpha \in \mathbf{CS}(X_k)$, $\boldsymbol{a},\boldsymbol{a}' \in \Delta_n$, $\alpha(\boldsymbol{a}') \in W_k$, $\delta(\boldsymbol{a}) = \delta(\boldsymbol{a}')$ and $w_{k-1}(\boldsymbol{a}) = w_{k-1}(\boldsymbol{a}')$, then $e_{\delta(\boldsymbol{a})}(\gamma_{\delta(\boldsymbol{a})}(\alpha(\boldsymbol{a}')),\boldsymbol{a}) \in W_k$;
3. $X_{k-1}|(\Delta_n \setminus \Delta_{k-2}) \subset \operatorname{int} W_k|(\Delta_n \setminus \Delta_{k-2})$ in $X_k|(\Delta_n \setminus \Delta_{k-2})$; and
4. there exists an N^n_0-map $H_k\colon I \times W_k \to X_k$ over Δ_n such that
   (i) $H_k(0,\langle g,\boldsymbol{a}\rangle) = \langle g,\boldsymbol{a}\rangle$ for all $\langle g,\boldsymbol{a}\rangle$;
   (ii) $H_k(t,\langle g,\boldsymbol{a}\rangle) = \langle g,\boldsymbol{a}\rangle$ for all t and all $\langle g,\boldsymbol{a}\rangle \in X_{k-1}$; and
   (iii) $H_k(1,\langle g,\boldsymbol{a}\rangle) \in X_{k-1}$ for all $\langle g,\boldsymbol{a}\rangle$.

Since W_k satisfies the assumptions of Proposition 6.8, there exists a continuous N^n_0-equivariant map $R_k\colon I \times \mathbf{CS}(X_k) \times \Delta_n \to X_k$ over Δ_n such that

1. $R_k(0,\alpha,\boldsymbol{a}) = \alpha(\boldsymbol{a})$ for all α and $\boldsymbol{a}$;
2. $R_k(t,\alpha,\boldsymbol{a}) = \alpha(\boldsymbol{a})$ for all t, all $\alpha \in \mathbf{CS}(X_{k-1})$ and all $\boldsymbol{a}$; and
3. $R_k(1,\alpha,\boldsymbol{a}) \in W_k$ for all α and $\boldsymbol{a}$.

Since $R_k(1,\alpha,\boldsymbol{a}) \in W_k$, one may join R_k and H_k to obtain

$$Q_k\colon I \times \mathbf{CS}(X_k) \times \Delta_n \to X_k,$$

where

$$Q_k(t,\alpha,\boldsymbol{a}) = \begin{cases} R_k(2t,\alpha,\boldsymbol{a}), & \text{when } 0 \le t \le 1/2; \text{ and} \\ H_k(2t-1, R_k(1,\alpha,\boldsymbol{a})), & \text{when } 1/2 \le t \le 1. \end{cases}$$

Both parts are continuous N^n_0-equivariant maps over Δ_n and they coincide when $t = 1/2$. Hence Q_k is a well-defined N^n_0-equivariant map over Δ_n.

The map Q_k has the following properties:

1. $Q_k(0,\alpha,\boldsymbol{a}) = R_k(0,\alpha,\boldsymbol{a}) = \alpha(\boldsymbol{a})$ for all α and $\boldsymbol{a}$;
2. $Q_k(t,\alpha,\boldsymbol{a}) = \alpha(\boldsymbol{a})$ for all t, all $\alpha \in \mathbf{CS}(X_{k-1})$ and all $\boldsymbol{a}$. For if $t \leq 1/2$, then $Q_k(t,\alpha,\boldsymbol{a}) = R_k(2t,\alpha,\boldsymbol{a}) = \alpha(\boldsymbol{a})$, and if $1/2 \leq t$, then $R_k(1,\alpha,\boldsymbol{a}) = \alpha(\boldsymbol{a}) \in X_{k-1}$ so that $Q_k(t,\alpha,\boldsymbol{a}) = H_k(2t-1, R_k(1,\alpha,\boldsymbol{a})) = \alpha(\boldsymbol{a})$, by Property (ii) of H_k.
3. $Q_k(1,\alpha,\boldsymbol{a}) = H_k(1, R_k(1,\alpha,\boldsymbol{a})) \in X_{k-1}$, by Property (iii) of H_k.

The adjoint of Q_k is continuous by Proposition 3.4 and it gives a strong N_0^n-deformation of $\mathbf{CS}(X_k)$ onto its N_0^n-invariant subspace $\mathbf{CS}(X_{k-1})$. □

The previous result can be used as the inductive step in the following result.

Corollary 6.10. *Suppose that for each $1 \leq k \leq n$ the subgroup N_{k-1}^n is a strong equivariant neighbourhood deformation retract of N_k^n. Then $\mathbf{CS}(X_0)$ is a strong N_0^n-deformation retract of $\mathbf{CS}(X_n)$.* □

7. The orbit $N_0^n \cdot \eta$ is a strong N_0^n-deformation retract of $\mathbf{CS}(X_0)$

Recall our standing assumptions from the beginning of Section 6.

Our main goal here in Part II has been to show that the orbit $N_0^n \cdot \eta$, considered as a subgroup, is a strong equivariant deformation retract of the group $\mathbf{CS}(X_n)$ when G is a metrisable group and for every $1 \leq k \leq n$ the subgroup N_{k-1}^n is a strong equivariant neighbourhood deformation retract of the group N_k^n, see Theorem 7.2.

The hard part was to show that $\mathbf{CS}(X_0)$ is a strong N_0^n-deformation retract of $\mathbf{CS}(X_n)$, which we did in Section 6. What remains is to show that the orbit $N_0^n \cdot \eta$ is a strong N_0^n-deformation retract of $\mathbf{CS}(X_0)$, which can be done by an easy application of Proposition 4.4.

Theorem 7.1. *The orbit $N_0^n \cdot \eta$ is a strong N_0^n-deformation retract of $\mathbf{CS}(X_0)$.*

Proof. Observe first that as a consequence of Proposition 2.3, a cross-section $\alpha \in \mathbf{CS}(X_0)$ belongs to $N_0^n \cdot \eta$ if and only if for all $\boldsymbol{a}$ one has that

$$\gamma_{\delta(\boldsymbol{a})}(\alpha(\boldsymbol{a})) = \gamma_{\delta(\boldsymbol{a})}(\alpha(\mathbf{e}_n)).$$

Let $H\colon I \times \Delta_n \to \Delta_n$ be a deformation of Δ_n which satisfies the following properties.

1. $H(0,\boldsymbol{a}) = \boldsymbol{a}$ for all $\boldsymbol{a}$;
2. $H(1,\boldsymbol{a}) = \mathbf{e}_n$ for all $\boldsymbol{a}$; and
3. $\delta(H(t,\boldsymbol{a})) \geq \delta(\boldsymbol{a})$ for all $\boldsymbol{a}$ and all t.

One may, for example, choose

$$H(t,\boldsymbol{a})_l = \begin{cases} (1-t)a_l, & \text{when } 0 \leq l \leq n-1; \text{ and} \\ a_n + t(a_0 + \cdots + a_{n-1}), & \text{when } l = n. \end{cases}$$

Only Property 3 requires some thought. Let $l = \delta(\boldsymbol{a})$. Then $a_l > 0$ and $a_{l'} = 0$ for all $l' > l$. If $l \leq n-1$ and $t < 1$, then $H(t,\boldsymbol{a})_l = (1-t)a_l > 0$, so that

$\delta(H(t,\boldsymbol{a})) \geq l$. If $l \leq n-1$ and $t=1$, then $H(t,\boldsymbol{a})_n = a_n + 1(a_0 + \cdots + a_{n-1}) = 1$, so that $\delta(H(t,\boldsymbol{a})) = n \geq l$. If $l = n$, then $H(t,\boldsymbol{a})_n = a_n + t(a_0 + \cdots + a_{n-1}) \geq a_n > 0$, so that $\delta(H(t,\boldsymbol{a})) = n = l$.

Proposition 4.4 shows that

$$\begin{array}{rccc} E\colon & I \times \mathbf{CS}(X_0) \times \Delta_n & \to & X_0 \\ & (t,\alpha,\boldsymbol{a}) & \mapsto & e_{\delta(\boldsymbol{a})}(\gamma_{\delta(\boldsymbol{a})}(\alpha(H(t,\boldsymbol{a}))),\boldsymbol{a}) \end{array}$$

is a well-defined continuous N_0^n-equivariant map over Δ_n.

The map E has the following properties:

1. For all α and all $\boldsymbol{a}$ it holds that
$$\gamma_{\delta(\boldsymbol{a})}(E(0,\alpha,\boldsymbol{a})) = \gamma_{\delta(\boldsymbol{a})}(\alpha(H(0,\boldsymbol{a}))) = \gamma_{\delta(\boldsymbol{a})}(\alpha(\boldsymbol{a})),$$
and so $E(0,\alpha,\boldsymbol{a}) = \alpha(\boldsymbol{a})$.
2. For all t, all $\alpha \in N_0^n \cdot \eta$ and all $\boldsymbol{a}$ one has
$$\gamma_{\delta(\boldsymbol{a})}(E(t,\alpha,\boldsymbol{a})) = \gamma_{\delta(\boldsymbol{a})}(\alpha(H(t,\boldsymbol{a}))) = \gamma_{\delta(\boldsymbol{a})}(\alpha(\mathbf{e}_n)) = \gamma_{\delta(\boldsymbol{a})}(\alpha(\boldsymbol{a})),$$
and so $E(t,\alpha,\boldsymbol{a}) = \alpha(\boldsymbol{a})$ according to Proposition 2.3. Note here that $\delta(H(t,\boldsymbol{a})) \geq \delta(\boldsymbol{a})$.
3. For all α and all $\boldsymbol{a}$ it holds that
$$\gamma_{\delta(\boldsymbol{a})}(E(1,\alpha,\boldsymbol{a})) = \gamma_{\delta(\boldsymbol{a})}(\alpha(H(1,\boldsymbol{a}))) = \gamma_{\delta(\boldsymbol{a})}(\alpha(\mathbf{e}_n)) = \gamma_{\delta(\boldsymbol{a})}(E(1,\alpha,\mathbf{e}_n)).$$

The adjoint $\hat{E}$ of E is continuous by Proposition 3.4 and it gives an N_0^n-equivariant map

$$\hat{E}\colon I \times \mathbf{CS}(X_0) \to \mathbf{CS}(X_0)$$

such that

1. $\hat{E}(0,\alpha) = \alpha$ for all α;
2. $\hat{E}(t,\alpha) = \alpha$ for all t and all $\alpha \in N_0^n \cdot \eta$; and
3. $\hat{E}(1,\alpha) \in N_0^n \cdot \eta$ for all α. □

Theorem 7.2. *Suppose that for each $1 \leq k \leq n$ the subgroup N_{k-1}^n is a strong equivariant neighbourhood deformation retract of N_k^n. Then the orbit $N_0^n \cdot \eta$, considered as a subgroup, is a strong equivariant deformation retract of* $\mathbf{CS}(X_n)$.

Proof. Since for each $1 \leq k \leq n$ the subgroup N_{k-1}^n is a strong equivariant neighbourhood deformation retract of N_k^n, it follows from Corollary 6.10 that the N_0^n-subspace $\mathbf{CS}(X_0)$ is a strong N_0^n-deformation retract of $\mathbf{CS}(X_n)$. Furthermore, the orbit $N_0^n \cdot \eta$ is a strong N_0^n-deformation retract of $\mathbf{CS}(X_0)$ by Theorem 7.1. Hence the orbit $N_0^n \cdot \eta$ is a strong N_0^n-deformation retract of $\mathbf{CS}(X_n)$. But we know that the N_0^n-action on $\mathbf{CS}(X_n)$ factors through the homomorphism $\lambda\colon N_0^n \to \mathbf{CS}(X_n)$. It follows that the orbit $N_0^n \cdot \eta$, considered as a subgroup, is a strong equivariant deformation retract of $\mathbf{CS}(X_n)$. □

By a *Lie group* we here mean a real and finite dimensional Lie group. Well-known results concerning Lie groups together with the result above imply the following.

Theorem 7.3. *Let G be a Lie group and let $K_0 \supset \cdots \supset K_n$ be compact subgroups. Then the orbit $N_0^n \cdot \eta$, considered as a subgroup, is a strong equivariant deformation retract of $\mathbf{CS}(X_n)$.*

Proof. Since G is a Lie group, it is metrisable, see [Bou72], Proposition III.1.1. Then the groups N_k^n are closed subgroups of G, and hence Lie groups, see [Che46], p. 135. Since N_{k-1}^n is a closed subgroup of the Lie group N_k^n, there exists a neighbourhood V of $N_{k-1}^n e$ in $N_{k-1}^n \backslash N_k^n$ and a cross-section $\sigma \colon V \to N_k^n$ of the canonical map $\pi \colon N_k^n \to N_{k-1}^n \backslash N_k^n$ such that V is homeomorphic to a neighbourhood of the origin in some euclidean space, see [Hel78], Lemma II.4.1. It follows that $N_{k-1}^n \backslash N_k^n$ is locally contractible. Theorem 5.1 then shows that N_{k-1}^n is a strong equivariant neighbourhood deformation retract of N_k^n. □

Part III. The group $\mathbf{Homeo}_{\Delta_n}^G(X)$ and its strong equivariant deformation retract $\psi(N_0^n)$

Recall our standing assumptions from the beginning of Section 6.

In Part II it was shown that $\mathbf{CS}(X_n)$ is a topological group that has a natural N_0^n-action. It was also shown that the N_0^n-orbit of the neutral cross-section η is a subgroup of $\mathbf{CS}(X_n)$ and that $N_0^n \cdot \eta$ is an equivariant deformation retract of $\mathbf{CS}(X_n)$. We now go on to discuss the group $\mathbf{Homeo}_{\Delta_n}^G(X)$ of orbit preserving G-homeomorphisms of the equivariant simplex X.

In Section 8 we will see that the group $\mathbf{CS}(X_n)$ can be identified, both as a group and as a topological space, with the group $\mathbf{Homeo}_{\Delta_n}^G(X)$ of G-equivariant homeomorphisms of X over Δ_n. Consequently, $\mathbf{Homeo}_{\Delta_n}^G(X)$ equipped with the compact-open topology is a topological group. The construction makes use of an isotropy characterisation of $\mathbf{CS}(X_n)$, which we give in Section 8.1.

Recall that the equivariant simplex X also exhibits a natural *right* action of the group N_0^n, see Section 2.2. In Section 8.3 we shall see that the orbit $N_0^n \cdot \eta$ corresponds to the group $\psi(N_0^n)$ of right translations of X, which is a subgroup of $\mathbf{Homeo}_{\Delta_n}^G(X)$.

The thesis is concluded with Section 9, where we discuss some implications and applications of the main results.

8. The correspondence between $\mathbf{CS}(X_n)$ and $\mathbf{Homeo}_{\Delta_n}^G(X)$

8.1. Isotropy characterisation of $\mathbf{CS}(X_n)$

We recall from Section 4.2 that

$$X_n = \{ \langle g, \boldsymbol{a} \rangle \in X \mid gK_{\delta(\boldsymbol{a})} \in N_{\delta(\boldsymbol{a})}^n / K_{\delta(\boldsymbol{a})} \}.$$

Observe that a point $x = \langle g, \boldsymbol{a} \rangle \in X_n$ has the group $K_{\delta(\boldsymbol{a})}$ as its isotropy subgroup G_x. However, there may exist points $x = \langle g, \boldsymbol{a} \rangle \in X$ with $G_x = K_{\delta(\boldsymbol{a})}$ which do not belong to X_n. In this respect, the result in Proposition 8.2 is non-trivial. Before we proceed to prove Proposition 8.2, let us point up a simple result, which is pure algebra.

Lemma 8.1. *Let $1 \leq k \leq n$ and let $g \in G$ be such that*

$$gK_{k-1} \in N_{k-1}/K_{k-1} \cap N_k^n/K_{k-1}.$$

Then $gK_{k-1} \in N_{k-1}^n/K_{k-1}$.

Proof. Since $gK_{k-1} \in N_{k-1}/K_{k-1}$ and since $K_{k-1} \subset N_{k-1}$, it follows that $g \in N_{k-1}$. Since $gK_{k-1} \in N_k^n/K_{k-1}$, there exists a $g' \in N_k^n$ such that $g'^{-1}g \in K_{k-1} \subset N_{k-1}$. Consequently, $g'^{-1} \in N_{k-1}$. Hence $g' \in N_{k-1}^n$, from which the claim follows. □

Proposition 8.2. *Let $\alpha\colon \Delta_n \to X$ be a cross-section such that $G_{\alpha(\boldsymbol{a})} = K_{\delta(\boldsymbol{a})}$ for all $\boldsymbol{a} \in \Delta_n$. Then $\alpha(\Delta_n) \subset X_n$.*

Proof. Recall that $G_{\alpha(\boldsymbol{a})} = K_{\delta(\boldsymbol{a})}$ is equivalent to

$$\gamma_{\delta(\boldsymbol{a})}(\alpha(\boldsymbol{a})) \in N_{\delta(\boldsymbol{a})}/K_{\delta(\boldsymbol{a})}.$$

It needs to be shown that

$$\gamma_{\delta(\boldsymbol{a})}(\alpha(\boldsymbol{a})) \in N_{\delta(\boldsymbol{a})}^n/K_{\delta(\boldsymbol{a})}$$

for all $\boldsymbol{a} \in \Delta_n$. We use descending induction on k to show that

$$\gamma_k(\alpha(\boldsymbol{a})) \in N_k^n/K_k$$

holds for all $\boldsymbol{a} \in \Delta_k \setminus \Delta_{k-1}$. If $k = n$, this is clear. Now assume that it holds for a certain $k \geq 1$.

Let

$$F_{k-1} = \{\ \boldsymbol{a} \in \Delta_n \setminus \Delta_{k-2} \mid \gamma_{k-1}(\alpha(\boldsymbol{a})) \in N_k^n/K_{k-1}\ \}.$$

Since $\gamma_{k-1} \circ \alpha$ is continuous and since N_k^n/K_{k-1} is closed in G/K_{k-1}, F_{k-1} is closed in $\Delta_n \setminus \Delta_{k-2}$. If $\boldsymbol{a} \in \Delta_k \setminus \Delta_{k-1}$, then according to the induction assumption

$$\gamma_k(\alpha(\boldsymbol{a})) \subset N_k^n/K_k,$$

and consequently, $\gamma_{k-1}(\alpha(\boldsymbol{a})) \in N_k^n/K_{k-1}$. Hence F_{k-1} contains the set $\Delta_k \setminus \Delta_{k-1}$. Since F_{k-1} is closed in $\Delta_n \setminus \Delta_{k-2}$, and since $\Delta_{k-1} \setminus \Delta_{k-2} \subset \overline{\Delta_k \setminus \Delta_{k-1}}$, we see that F_{k-1} must also contain the set $\Delta_{k-1} \setminus \Delta_{k-2}$. This means that $\gamma_{k-1}(\alpha(\boldsymbol{a})) \in N_k^n/K_{k-1}$ holds for all $\boldsymbol{a} \in \Delta_{k-1} \setminus \Delta_{k-2}$. Since $\gamma_{k-1}(\alpha(\boldsymbol{a})) \in N_{k-1}/K_{k-1}$ holds by assumption for all $\boldsymbol{a} \in \Delta_{k-1} \setminus \Delta_{k-2}$, it follows from Lemma 8.1 that $\gamma_{k-1}(\alpha(\boldsymbol{a})) \in N_{k-1}^n/K_{k-1}$ holds for all $\boldsymbol{a} \in \Delta_{k-1} \setminus \Delta_{k-2}$. This completes the induction. □

Note that the converse statement is trivial: if $\alpha(\Delta_n) \subset X_n$, then $G_{\alpha(\boldsymbol{a})} = K_{\delta(\boldsymbol{a})}$ for all $\boldsymbol{a} \in \Delta_n$. As a result, we may characterise the set $\mathbf{CS}(X_n)$ as

$$\mathbf{CS}(X_n) = \{\ \alpha \in \mathbf{CS}(X) \mid G_{\alpha(\boldsymbol{a})} = K_{\delta(\boldsymbol{a})} \text{ for all } \boldsymbol{a} \in \Delta_n\ \}.$$

8.2. Cross-sectional behaviour of G-homeomorphisms of X over Δ_n

Let $f\colon X \to X$ be a *G-map over* Δ_n. It follows that f is surjective. Suppose that f is in addition *isovariant*, i.e., $G_{f(x)} = G_x$ for all x. It now follows that f is bijective and indeed a *G-homeomorphism.* Namely, if $f(x) = f(y)$, then $y = gx$ for some $g \in G$ since f is a map over Δ_n. But then $f(x) = f(gx) = gf(x)$, and hence $g \in G_{f(x)} = G_x$, so it follows that $y = gx = x$. Hence f is bijective. We will soon see that the inverse of f is also continuous.

First, note that since f is a map over Δ_n, it follows that $f\eta$ is a cross-section of X. Moreover, if f is bijective, it is clearly isovariant, and hence for each $\boldsymbol{a} \in \Delta_n$ it holds that

$$G_{f\eta(\boldsymbol{a})} = G_{\eta(\boldsymbol{a})} = K_{\delta(\boldsymbol{a})}.$$

Therefore, by Proposition 8.2, $f\eta$ is a cross-section of X_n. Let us temporarily consider the set $\mathbf{Bij}^G_{\Delta_n}(X)$ of bijective G-maps over Δ_n. Giving this the compact-open topology, we obtain

$$\eta^*\colon \mathbf{Bij}^G_{\Delta_n}(X) \to \mathbf{CS}(X_n),$$

which is continuous by Proposition 3.2.

In the above setting η^* is injective. For if f and h are two G-maps over Δ_n such that $f\eta = h\eta$, then

$$f(\langle g, \boldsymbol{a}\rangle) = f(g\eta(\boldsymbol{a})) = gf\eta(\boldsymbol{a}) = gh\eta(\boldsymbol{a}) = h(g\eta(\boldsymbol{a})) = h(\langle g, \boldsymbol{a}\rangle),$$

which implies that $f = h$.

We shall now show that η^* is an antihomomorphism. Let h and f be two maps in $\mathbf{Bij}^G_{\Delta_n}(X)$. The claim is that

$$\eta^*(hf) = \eta^*(f) \star \eta^*(h).$$

So let $\boldsymbol{a} \in \Delta_n$ and choose representatives $l, l' \in N^n_{\delta(\boldsymbol{a})}$ such that

$$\begin{cases} f(\eta(\boldsymbol{a})) = \langle l, \boldsymbol{a}\rangle = l\eta(\boldsymbol{a}); \text{ and} \\ h(\eta(\boldsymbol{a})) = \langle l', \boldsymbol{a}\rangle = l'\eta(\boldsymbol{a}). \end{cases}$$

Since h is G-equivariant,

$$\eta^*(hf)(\boldsymbol{a}) = h(f(\eta(\boldsymbol{a}))) = h(l\eta(\boldsymbol{a})) = lh(\eta(\boldsymbol{a})) = ll'\eta(\boldsymbol{a}) = (\eta^*(f) \star \eta^*(h))(\boldsymbol{a}),$$

as required.

We shall now build an inverse for η^*. We define a function

$$\epsilon\colon \mathbf{CS}(X_n) \to \mathbf{Bij}^G_{\Delta_n}(X)$$

by setting

$$\epsilon(\alpha)(\langle g, \boldsymbol{a}\rangle) = g\alpha(\boldsymbol{a}).$$

Observe that

$$\langle g, \boldsymbol{a}\rangle = \langle g', \boldsymbol{a}\rangle \Leftrightarrow g'^{-1}g \in K_{\delta(\boldsymbol{a})} = G_{\alpha(\boldsymbol{a})} \Leftrightarrow g\alpha(\boldsymbol{a}) = g'\alpha(\boldsymbol{a}).$$

It follows that the value of $\epsilon(\alpha)$ at $\langle g, \boldsymbol{a}\rangle$ is a well-defined element of X and also that $\epsilon(\alpha)$ is injective. One also sees that $\epsilon(\alpha)$ satisfies

$$\epsilon(\alpha) \circ p = \Phi \circ (\mathrm{id} \times \alpha),$$

where Φ denotes the G-action on X. Since p is a quotient map, one sees straightaway that $\epsilon(\alpha)$ is a G-map over Δ_n, so $\epsilon(\alpha)$ belongs to $\mathbf{Bij}^G_{\Delta_n}(X)$.

We will next show that ϵ is the inverse of η^*. Note that

$$((\eta^* \circ \epsilon)(\alpha))(\boldsymbol{a}) = \epsilon(\alpha)(\langle e, \boldsymbol{a}\rangle) = \alpha(\boldsymbol{a}),$$

so that $\eta^* \circ \epsilon = \mathrm{id}$. But this suffices, since we already know that η^* is injective.

A priori, $\mathbf{Bij}^G_{\Delta_n}(X)$ is a monoid. But since η^* is a well-defined anti-isomorphism and since $\mathbf{CS}(X_n)$ is a group, it follows that $\mathbf{Bij}^G_{\Delta_n}(X)$ coincides with $\mathbf{Homeo}^G_{\Delta_n}(X)$.

We already know that η^* is continuous, but in fact it is a homeomorphism. Namely, let us consider the factorisation of $\epsilon(\alpha)$ given above. The right hand side is continuous in α since we may factor it as $\Phi_* \circ (\mathrm{id} \times \bullet)$, where

$$\begin{array}{rccc} \mathrm{id} \times \bullet\colon & \mathbf{CS}(X) & \to & \mathbf{map}(G \times \Delta_n, G \times X) \\ & \alpha & \mapsto & \mathrm{id} \times \alpha \end{array}$$

is the restricted product functor, the continuity of which is dealt with in Proposition 3.6, and where Φ_* is continuous by Proposition 3.1. Then again, the left hand side factors as $p^* \circ \epsilon$ and now since p is a proper surjective map, p^* is an embedding by Proposition 3.2. Thus ϵ is continuous.

The above discussion leads to the following theorem.

Theorem 8.3. *The maps*

$$\mathbf{CS}(X_n) \underset{\eta^*}{\overset{\epsilon}{\rightleftarrows}} \mathbf{Homeo}^G_{\Delta_n}(X)$$

are homeomorphisms and anti-isomorphisms of groups and they are inverses of each other. Consequently, $\mathbf{Homeo}^G_{\Delta_n}(X)$ *is a topological group.*

Proof. It remains to show that $\mathbf{Homeo}^G_{\Delta_n}(X)$ is a topological group. It was shown in Theorem 4.3 that $\mathbf{CS}(X_n)$ is a topological group with respect to multiplication of cross-sections.

To construct a topological isomorphism between $\mathbf{Homeo}^G_{\Delta_n}(X)$ and $\mathbf{CS}(X_n)$ one may pre- or postcompose η^* with the inverse-element map of the corresponding group. The resulting map is of course the same in both cases. Therefore, $\mathbf{Homeo}^G_{\Delta_n}(X)$ is a topological group with respect to composition of maps. □

8.3. The group $\psi(N_0^n)$ of right translations of X

In Section 2 we showed that besides having a left G-action $\Phi\colon G \times X \to X$, the equivariant simplex X also exhibits a natural *right* N_0^n-action

$$\begin{array}{rccc} \Psi\colon & X \times N_0^n & \to & X \\ & (\langle g, \boldsymbol{a}\rangle, n) & \mapsto & \langle gn, \boldsymbol{a}\rangle \end{array}.$$

Since Φ and Ψ commute and since Ψ preserves the G-orbits, the adjoint $\psi = \hat{\Psi}$ gives a canonical antihomomorphism

$$\psi\colon N_0^n \to \mathbf{Homeo}_{\Delta_n}^G(X).$$

Let us make a short remark on the relation between ψ and the homomorphism

$$\varphi\colon G \to \mathbf{Homeo}_{\Delta_n}(X),$$

which was defined in Section 4.1. Observe that for all $n \in N_0^n$ it holds that

$$\varphi(n) \circ \eta = n \cdot \eta = \psi(n) \circ \eta,$$

where $\varphi(n)$ is the left transformation of X that corresponds to n. However, it is *important to distinguish* $\varphi(n)$ *and* $\psi(n)$. Although both are homeomorphisms of X over Δ_n and though they coincide on the cross-sectional set $\eta(\Delta_n)$, the transformation $\varphi(g)$ need not be G-equivariant even for $g \in N_0^n$.

Recall that N_0^n acts on $\mathbf{CS}(X_n)$ through a proper homomorphism λ whose image is the orbit $N_0^n \cdot \eta$ and whose kernel is K_n. In Section 8.2 it was shown that the topological group $\mathbf{CS}(X_n)$ can be identified with the group $\mathbf{Homeo}_{\Delta_n}^G(X)$ using the anti-isomorphisms

$$\mathbf{CS}(X_n) \underset{\eta^*}{\overset{\epsilon}{\rightleftarrows}} \mathbf{Homeo}_{\Delta_n}^G(X),$$

see Theorem 8.3. Then $\epsilon \circ \lambda\colon N_0^n \to \mathbf{Homeo}_{\Delta_n}^G(X)$ is a proper antihomomorphism with kernel K_n and we claim that it is the same map as ψ. So let $n \in N_0^n$ and let $\langle g, \boldsymbol{a}\rangle \in X$. Now

$$(\epsilon \circ \lambda)(n)(\langle g, \boldsymbol{a}\rangle) = \epsilon(n \cdot \eta)(\langle g, \boldsymbol{a}\rangle) = g(n \cdot \eta)(\boldsymbol{a}) = gn\langle e, \boldsymbol{a}\rangle = \langle gn, \boldsymbol{a}\rangle = \psi(n)(\langle g, \boldsymbol{a}\rangle),$$

which shows that $(\epsilon \circ \lambda)(n) = \psi(n)$. Since ψ is a proper map and an antihomomorphism with kernel K_n, the image $\psi(N_0^n)$ is a closed subgroup of $\mathbf{Homeo}_{\Delta_n}^G(X)$, which is isomorphic to N_0^n/K_n.

The above also shows that the left action of $N_0^n \cdot \eta$ on $\mathbf{CS}(X_n)$ may be transformed into the right action

$$\begin{array}{rccc} \Psi\colon & \mathbf{Homeo}_{\Delta_n}^G(X) \times \psi(N_0^n) & \to & \mathbf{Homeo}_{\Delta_n}^G(X) \\ & (h, f) & \mapsto & hf, \end{array}$$

which we may regard as being the "same" action as the left action of $N_0^n \cdot \eta$ on $\mathbf{CS}(X_n)$.

We are now ready to restate our main results in a new form. The following corresponds to Theorem 7.2.

Theorem 8.4. *Suppose that for each $1 \leq k \leq n$ the subgroup N^n_{k-1} is a strong equivariant neighbourhood deformation retract of N^n_k. Then the subgroup $\psi(N^n_0)$ is a strong equivariant deformation retract of* $\mathbf{Homeo}^G_{\Delta_n}(X)$. □

For Lie groups we may restate Theorem 7.3.

Theorem 8.5. *Let G be a Lie group. Then the subgroup $\psi(N^n_0)$ is a strong equivariant deformation retract of* $\mathbf{Homeo}^G_{\Delta_n}(X)$. □

9. Implications and applications of the main results and further remarks

The main results of this thesis are Theorems 7.2, 7.3, 8.4 and 8.5, which can all be considered more or less as one and the same result. In Section 9.1 we give a brief discussion of some implications of the main results for the structure of $\mathbf{Homeo}^G_{\Delta_n}(X)$. In Section 9.2 we show how an established result in the literature, the Straightening Lemma, can be obtained as a special case of our main theorem. In Section 9.3 we point out a connection between Theorem 7.3 and another existing result in the literature which is used to prove the uniqueness of G-CW complex structures. As it happens, the proof of this result appears not to be correctly formulated. We will show that Theorem 7.3 provides a correct proof using a completely different approach.

In Section 9.4 we apply our main results to show that every G-homeomorphism $h\colon X \to X$ over Δ_n is homotopic through G-homeomorphisms over Δ_n to some element in the group $\psi(N^n_0)$ of right translations. We conclude the thesis in Section 9.5 by showing that Theorem 8.4 gives a non-trivial generalisation of Theorem 8.5.

9.1. Implications of the main results

In this section we point up a few simple implications of the main results for the structure of $\mathbf{Homeo}^G_{\Delta_n}(X)$. Similar results of course hold for $\mathbf{CS}(X_n)$.

Corollary 9.1. *Let X be as in Theorem 8.5. Then $\mathbf{Homeo}^G_{\Delta_n}(X)$ has the homotopy type of the Lie group N^n_0/K_n.*

Proof. According to Theorem 8.5, the subgroup $\psi(N^n_0)$ is a strong deformation retract of $\mathbf{Homeo}^G_{\Delta_n}(X)$, so $\mathbf{Homeo}^G_{\Delta_n}(X)$ has the same homotopy type as $\psi(N^n_0)$, see [Spa66], Lemma 1.4.4, for example. It was shown in Section 8.3 that $\psi(N^n_0)$ is homeomorphic to N^n_0/K_n. Since $N_n = N(K_n)$ is a closed subgroup of G, it is a Lie group by [Che46], p. 135. Similarly K_n is a closed normal subgroup of N_n. Hence the factor group N_n/K_n can be given the structure of a Lie group, see [Bou72], Proposition III.1.11(ii). Finally, observe that N^n_0/K_n is a closed subgroup of N_n/K_n. □

Corollary 9.2. *Let X be as in Theorem 8.4. Then*

1. $\psi(N_0^n)$ *has a global cross-section in* $\mathbf{Homeo}^G_{\Delta_n}(X)$*; and*
2. $\mathrm{id}\,\psi(N_0^n)$ *is a strong deformation retract of* $\mathbf{Homeo}^G_{\Delta_n}(X)/\psi(N_0^n)$.

Proof. This follows from Theorems 8.4 and 5.1 and Remark 5.2. □

Let us also note that it follows from Corollary 9.2, the Bundle structure theorem 7.4 and the Cross-section theorem 8.3 in [Ste51] that the canonical map

$$\pi\colon \mathbf{Homeo}^G_{\Delta_n}(X) \to \mathbf{Homeo}^G_{\Delta_n}(X)/\psi(N_0^n)$$

is a trivial bundle.

9.2. Applications to the Straightening Lemma

The purpose of the following exposition is to highlight a connection to an existing result in the literature rather than to promote a new proof of it.

Let G be a compact Lie group, let $H \subset K$ be closed subgroups and let $\pi\colon G/H \to G/K$ be the canonical map. The *mapping cylinder* M_π of π is the quotient space obtained from the topological sum of $G/H \times I$ and G/K by identifying $(gH, 1)$ with gK. The space M_π is in a natural way a *left* G-space and its orbit space can be identified with I. The group $N(H) \cap N(K)$ acting on G by multiplication from the right induces a *right* action on G/H and G/K, see Example 1.4. There is an induced right action of $N(H) \cap N(K)$ on M_π, whose ineffective kernel is H. The left and the right action commute and the right action induces a trivial action on I. Therefore, the adjoint of the right action gives an antihomomorphism $\hat{\psi}\colon N(H) \cap N(K) \to \mathbf{Homeo}^G_I(M_\pi)$, whose kernel is H. The *Straightening Lemma* states that when G is a compact Lie group, the subgroup corresponding to $(N(K) \cap N(H))/H$ is a strong deformation retract of $\mathbf{Homeo}^G_{\Delta_1}(M_\pi)$, see [Bre72], Lemma V.4.1.

We now proceed to show that M_π can in a natural way be considered as the one dimensional equivariant simplex $X = \Delta_1(G; K, H)$. While we are at it, let us do this for a general topological group G and compact subgroups K and H. We first present a short remark on attachings of spaces.

Lemma 9.3. *Let* $A \subset X$ *and let* $f\colon A \to Y$ *be a closed surjective map. Let* $X \cup_f Y$ *denote the space obtained by attaching* X *to* Y *on* A *using* f *and let* $q\colon X + Y \to X \cup_f Y$ *denote the identification. Then* $q\colon X \to X \cup_f Y$ *is a quotient map.*

Proof. The restricted map remains surjective. Let $F \subset X \cup_f Y$ be a set such that $q^{-1}(F) \cap X$ is closed in X. We claim that $q^{-1}(F)$ is closed in $X + Y$. It suffices to show that $q^{-1}(F) \cap Y$ is closed in Y. Note that $q^{-1}(F) \cap Y = f(q^{-1}(F) \cap A)$. To see this, recall that $f\colon A \to Y$ is surjective and note that $y \in q^{-1}(F) \cap Y$ if and only if there exists an $a \in A$ such that $f(a) = y$ and $q(a) \in F$, and this holds if and only if $y \in f(q^{-1}(F) \cap A)$. Now since $A \subset X$, $q^{-1}(F) \cap A$ is closed in A and since f is closed, $f(q^{-1}(F) \cap A)$ is closed in Y. □

We identify I with Δ_1 using the homeomorphism

$$\begin{array}{rccc} j\colon & I & \to & \Delta_1 \\ & t & \mapsto & (t, 1-t) \end{array}.$$

Proposition 9.4. *Let G be a topological group and let $H \subset K$ be compact subgroups. Let M_π denote the mapping cylinder of the canonical map $\pi\colon G/H \to G/K$ and let X denote the equivariant simplex $\Delta_1(G; K, H)$. Then there exists a homeomorphism $h\colon M_\pi \to X$ that is left G- and right $(N(K) \cap N(H))$-equivariant and induces $j\colon I \to \Delta_1$ between the G-orbit spaces.*

Proof. Let $q\colon G/H \times I + G/K \to M_\pi$ denote the identifying projection. Since $\pi_K\colon G \to G/K$ is a proper map by Proposition 1.11 and since $\pi_K = \pi\pi_H$, $\pi\colon G/H \to G/K$ is a proper map by [Bou66], Proposition I.10.5b, and π is surjective. Therefore, the restricted map $q\colon G/H \times I \to M_\pi$ is a quotient map according to Lemma 9.3.

Consider the maps

$$f = q \circ (\pi_H \times \mathrm{id})\colon G \times I \to M_\pi$$

and

$$f' = p \circ (\mathrm{id} \times j)\colon G \times I \to X.$$

Since $\pi_H\colon G \to G/H$ is a proper map by Proposition 1.11, the maps given above are quotient maps. They have exactly the same fibres and hence as in Example 1.6, there exists a homeomorphism $h\colon M_\pi \to X$ satisfying $hf = f'$. Since both the maps given above are left G-equivariant and right $(N(K) \cap N(H))$-equivariant, so must be the induced map h. □

Conjugation using h^{-1} yields an isomorphism

$$h^{-1}{}_*h^*\colon \mathbf{Homeo}^G_{\Delta_1}(X) \to \mathbf{Homeo}^G_I(M_\pi)$$

and it is a trivial matter to see that $\psi(N_0^n)$ corresponds to $(N(K) \cap N(H))/H$ considered as a subgroup of $\mathbf{Homeo}^G_I(M_\pi)$. Now, if G is a Lie group, we have seen in Theorem 8.5 that $\psi(N_0^n)$ is a strong deformation retract of $\mathbf{Homeo}^G_{\Delta_1}(X)$. Hence we see that the Straightening Lemma is a simple special case of Theorem 8.5 and the assumption that G is compact is unnecessary.

9.3. Applications to uniqueness of G-CW complex structures

The first formulation of the main result of this thesis, Theorem 7.3, concerns deformations of function spaces of cross-sections. By a simple application of the evaluation map we obtain Proposition 9.5, which concerns homotopies of single cross-sections. For G a compact Lie group, this result is found in the literature, see [MS87], Lemma 4.4. However, it seems that the proof there is not correctly formulated, as Sören Illman pointed out to the present author. Indeed, it is shown in Example 9.6 that the choice of the group element g is not correct in the proof found in [MS87].

Proposition 9.5. *Let X be an equivariant simplex as in Theorem 7.2 (or 7.3). Let α and β be two cross-sections of X such that for all $\boldsymbol{a} \in \Delta_n$*

$$G_{\alpha(\boldsymbol{a})} = K_{\delta(\boldsymbol{a})} = G_{\beta(\boldsymbol{a})}.$$

Then there exists a $g \in G$ and a homotopy $S\colon I \times \Delta_n \to X$ over Δ_n such that

1. *$S(0, \boldsymbol{a}) = \alpha(\boldsymbol{a})$ for all $\boldsymbol{a}$;*
2. *$S(1, \boldsymbol{a}) = g\beta(\boldsymbol{a})$ for all $\boldsymbol{a}$; and*
3. *$G_{S(t,\boldsymbol{a})} = K_{\delta(\boldsymbol{a})}$ for all t and all $\boldsymbol{a}$.*

Moreover, any such g has to belong to N_0^n.

Proof. Notice first that by Proposition 8.2 both α and β are cross-sections of X_n. Since $\mathbf{CS}(X_n)$ is a topological group, we may consider $\alpha \star \beta^{-1}$. Let $H\colon I \times \mathbf{CS}(X_n) \to \mathbf{CS}(X_n)$ be the deformation of Theorem 7.2. Recall that

1. $H(0, \alpha \star \beta^{-1}) = \alpha \star \beta^{-1}$; and
2. $H(1, \alpha \star \beta^{-1}) = g \cdot \eta$ for some $g \in N_0^n$.

Let us define a map

$$\begin{array}{rccc} S\colon & I \times \Delta_n & \to & X_n \\ & (t, \boldsymbol{a}) & \mapsto & \mathrm{e}(H(t, \alpha \star \beta^{-1}) \star \beta, \boldsymbol{a}) \end{array},$$

where $\mathrm{e}\colon \mathbf{CS}(X_n) \times \Delta_n \to X_n$ is the evaluation map. Since $(\alpha \star \beta^{-1}) \star \beta = \alpha$ and since $(g \cdot \eta) \star \beta = g \cdot \beta$, we see that S has the required properties.

Since it follows from 3 that both β and $g \cdot \beta$ belong to $\mathbf{CS}(X_n)$, we conclude that $g \cdot \eta = (g \cdot \beta) \star \beta^{-1}$ also belongs to $\mathbf{CS}(X_n)$. It follows that for all $\boldsymbol{a} \in \Delta_n$

$$gK_{\delta(\boldsymbol{a})}g^{-1} = gG_{\langle e,\boldsymbol{a}\rangle}g^{-1} = G_{\langle g,\boldsymbol{a}\rangle} = K_{\delta(\boldsymbol{a})}.$$

Hence $g \in N_0^n$. □

In the proof of Lemma 4.4 in [MS87] the group element g is constructed inductively, and for $n = 0$ it belongs to $N_0^n = N(K_0)$. What shows that the process does not yield a satisfactory element g is that the proposed homotopy constructed in [MS87] would hold $\alpha(\mathbf{e}_n)$ fixed, implying that $g\beta(\mathbf{e}_n) = \alpha(\mathbf{e}_n)$. The following example shows that this violates the constraint that by Proposition 9.5 the element g should also belong to N_0^n.

Example 9.6. Let G be a connected Lie group and let $K \subset G$ be a compact subgroup such that $N(K) \neq G$. An example where G is *compact* is given by $G = \mathbf{SO}_3(\mathbb{R})$ and $K = \mathbf{SO}_2(\mathbb{R})$. Let us consider $X = \Delta_1(G; K, \{e\})$.

Since G is a Lie group and since $N(K)$ is a proper closed subgroup, there exists a neighbourhood V of $N(K)e$ in $N(K)\backslash G$ and a cross-section $\sigma\colon V \to G$ of the canonical map $\pi\colon G \to N(K)\backslash G$ such that V is homeomorphic to a neighbourhood of the origin in some euclidean space of positive dimension, see [Hel78], Lemma II.4.1. Moreover, we may assume that the origin corresponds to the point $N(K)e$. Hence V contains a path $\gamma\colon I \to V$ such that $\gamma(0) \neq N(K)e$ and $\gamma(1) = N(K)e$. But then $\alpha' = \sigma\gamma\colon I \to G$ is a path such that $\alpha'(0) \notin N(K)$ and $\alpha'(1) \in N(K)$.

Let $j\colon I \to \Delta_1$ be the homeomorphism that appeared in Proposition 9.4. Then

$$\begin{array}{rccc} \alpha\colon & \Delta_1 & \to & X_1 \\ & \boldsymbol{a} & \mapsto & \langle \alpha'(j^{-1}(\boldsymbol{a})), \boldsymbol{a}\rangle \end{array}$$

is a cross-section of X_1. There does not exist an element g in G such that $g\eta(\mathbf{e}_1) = \langle g, \mathbf{e}_1\rangle = \alpha(\mathbf{e}_1)$ other than $\alpha'(0)$, which does not belong to $N_0^1 = N(K)$. Hence, even though we have seen in Proposition 9.5 that there exists a $g \in G$ such that α and $g \cdot \eta$ are homotopic as maps to X_1, they are not homotopic rel$\{\mathbf{e}_1\}$, although this would seem to be the case assuming the proof in [MS87] was correct. □

The choice of the group element g does not seem to be reflected in the uniqueness results of G-CW complexes. Therefore, one may assume that the uniqueness results in [MS87] are otherwise well founded; see, however, remarks at the beginning of Section 5 of [Ill00]. That the group element g belongs to N_0^n may indeed give new information about G-CW structures and related concepts.

9.4. Applications to homotopies of single maps

In Section 9.3 we dealt with isotropy structure preserving homotopies of single cross-sections. In this section we treat in an analogous fashion G-homotopy classes of single G-homeomorphisms $h\colon X \to X$ over Δ_n. In fact, for G a Lie group, this result was the original research question proposed to the author by Sören Illman.

Let X be an equivariant simplex as in Theorem 8.4. Since X is not necessarily locally compact, it is not possible to quote Proposition 3.3 for the continuity of the evaluation map.

Lemma 9.7. *The evaluation*

$$\begin{array}{rccc} \mathrm{e}\colon & \mathbf{Homeo}^G_{\Delta_n}(X) \times X & \to & X \\ & (f, x) & \mapsto & f(x) \end{array}$$

is continuous.

Proof. The map p is a proper surjection and thus the map

$$\mathrm{id} \times p\colon \mathbf{Homeo}^G_{\Delta_n}(X) \times G \times \Delta_n \to \mathbf{Homeo}^G_{\Delta_n}(X) \times X$$

is a closed surjective map, and more importantly, a quotient map. Now, for a suitably chosen permutation σ of the three factors it holds that

$$\mathrm{e} \circ (\mathrm{id} \times p) \circ \sigma = \Phi \circ (\mathrm{id}_G \times \mathrm{e} \circ (\mathrm{id} \times \eta)),$$

as can be verified by performing the calculation

$$\mathrm{e}(f, p(g, \boldsymbol{a})) = \mathrm{e}(f, g\eta(\boldsymbol{a})) = g\,\mathrm{e}(f, \eta(\boldsymbol{a})).$$

We conclude that e is continuous if and only if $\mathrm{e} \circ (\mathrm{id} \times \eta)$ is continuous.

The function

$$\begin{array}{rccc} \mathrm{e}'\colon & \mathbf{CS}(X_n) \times \Delta_n & \to & X_n \\ & (\alpha, \boldsymbol{a}) & \mapsto & \alpha(\boldsymbol{a}) \end{array}$$

is continuous by Proposition 3.3. But now $\mathrm{e} \circ (\mathrm{id} \times \eta) = \mathrm{e}' \circ (\eta^* \times \mathrm{id})$ and $\eta^* \times \mathrm{id}$ is a homeomorphism by Theorem 8.3. Therefore, e is continuous. □

Theorem 9.8. *Let X be an equivariant simplex as in Theorem 8.4 (or 8.5) and let $h\colon X \to X$ be a G-homeomorphism over Δ_n. Then h is homotopic through G-homeomorphisms over Δ_n to some element in the group $\psi(N_0^n)$.*

Proof. Let $D\colon I \times \mathbf{Homeo}_{\Delta_n}^G(X) \to \mathbf{Homeo}_{\Delta_n}^G(X)$ be the deformation of the group $\mathbf{Homeo}_{\Delta_n}^G(X)$ to its subgroup $\psi(N_0^n)$ given in Theorem 8.4. Let $\mathrm{e}\colon \mathbf{Homeo}_{\Delta_n}^G(X) \times X \to X$ be the evaluation map of Lemma 9.7. Recall that

1. $D(0,h) = h$; and
2. $D(1,h) \in \psi(N_0^n)$.

Then

$$\begin{aligned} H\colon \quad I \times X &\to X \\ (t,x) &\mapsto \mathrm{e}(D(t,h),x) \end{aligned}$$

is the desired G-homotopy over Δ_n. □

9.5. A non-Lie group example

The following example shows that Theorem 8.4 gives a non-trivial generalisation of Theorem 8.5.

Example 9.9. The aim is to construct a metrisable group G which is not locally compact and which has $n+1$ different compact subgroups $K_0 \supset \cdots \supset K_n$ such that for all $1 \le k \le n$ the subgroup N_{k-1}^n is an equivariant neighbourhood deformation retract of N_k^n.

Let us first show that it is sufficient to find a metrisable group G' which is not locally compact and which contains a non-trivial compact subgroup K' such that $N(K')$ is an equivariant neighbourhood deformation retract of G'.

Namely, let G denote the $(n+1)$-fold direct product of G', and for all $0 \le i \le n$ define

$$K_i = \underbrace{\{e\} \times \cdots \times \{e\}}_{i} \times \underbrace{K' \times \cdots \times K'}_{n+1-i}.$$

Notice that

$$N(K_i) = \underbrace{G' \times \cdots \times G'}_{i} \times \underbrace{N(K') \times \cdots \times N(K')}_{n+1-i}.$$

Observe that $N(K_{i-1}) \subset N(K_i)$ for all $1 < i \le n$ and hence $N_k^n = N(K_k)$.

Suppose that $N(K')$ has an open $N(K')$-invariant neighbourhood V in G' and an $N(K')$-equivariant map $H\colon I \times V \to G'$ such that $H(0,g) = g$ and $H(1,g) \in N(K')$ for all g. Then $U = {\mathrm{pr}_k}^{-1}(V)$ is an open N_{k-1}^n-invariant neighbourhood of N_{k-1}^n in

N^n_k and there exists an N^n_{k-1}-equivariant map $D_k\colon I \times U \to N^n_k$ given by

$$D_k(t, (g_1, \ldots, g_{n+1}))_i = \begin{cases} g_i, & \text{when } i \neq k; \text{ and} \\ H(t, g_k), & \text{when } i = k. \end{cases}$$

We have $D_k(0, g) = g$ and $D_k(1, g) \in N^n_{k-1}$ for all g. This shows that N^n_{k-1} is a strong neighbourhood deformation retract of N^n_k.

Let us now show that such G' and K' do exist. Let G' denote the group $\mathbf{Homeo}_+(S^1)$ of orientation preserving (i.e., degree 1) homeomorphisms of the circle and let K' denote the embedded S^1 in G', that is, the group of rotations of the circle S^1. Since S^1 is locally compact, locally connected and Hausdorff, G' is a topological group, see [Are46], Theorem 4. Since S^1 is a compact metric space, G' is metrisable, see [Dug66], XII.8.2(3). However, G' is not a Lie group for it is not even locally compact.

Since our task is to show that $N(K')$ is an $N(K')$-equivariant neighbourhood deformation retract of G', we should first identify the group $N(K')$. As it happens, $N(K') = K'$. Namely, let $f \in N(K')$. Then $k \mapsto f^{-1}kf$ is an automorphism of the Lie group K'. However, K' has only two continuous automorphisms, $k \mapsto k$ and $k \mapsto k^{-1}$. These yield in complex number presentation formulas $f(z) = zf(1)$ and $f(z) = \bar{z}f(1)$, respectively. Since f is orientation preserving, the latter is impossible, and hence f itself is a rotation.

What remains is to construct a strong S^1-deformation retract of the group $\mathbf{Homeo}_+(S^1)$ of orientation preserving homeomorphisms of S^1 onto its subgroup S^1 of rotations. Let $p\colon \mathbb{R} \to \mathbb{R}/\mathbb{Z} = S^1$ denote the covering projection $\phi \mapsto e^{2\pi i\phi}$. To every homeomorphism $f \in \mathbf{Homeo}_+(S^1)$ there corresponds a family of homeomorphisms $\tilde{f}\colon \mathbb{R} \to \mathbb{R}$ which have the property that $p\tilde{f} = fp$. The expression

$$p\Big(\int_{[0,1]} (\tilde{f}(\phi) - \phi)d\phi\Big)$$

is independent of the choice of a lifting since for any two liftings $\tilde{f}$ and $\tilde{f}'$, the difference $\tilde{f} - \tilde{f}'$ is an integer valued constant function. Hence, the expression gives us a function

$$r\colon \mathbf{Homeo}_+(S^1) \to S^1,$$

which represents the *average rotation angle* of f.

Since S^1 is a compact metric space, the compact-open topology on $\mathbf{Homeo}_+(S^1)$ is given by the supremum metric. For any sequence of maps f_n converging to f in $\mathbf{Homeo}_+(S^1)$ it is possible to choose liftings $\tilde{f}_n$ and $\tilde{f}$ in such a way that

$$\sup_{\phi \in [0,1]} |\tilde{f}_n(\phi) - \tilde{f}(\phi)| \to 0.$$

This, together with basic properties of the integral, shows us that r is a continuous function.

Let $\rho\colon S^1 \to S^1$ be a rotation by an angle of $\rho \cdot 2\pi$ and let $f \in \mathbf{Homeo}_+(S^1)$. If $\tilde{f}$ is a lifting of f, then $\tilde{f}+\rho$ if a lifting of $\rho \circ f$. This shows that r is S^1-equivariant. Moreover, since $r(\mathrm{id}) = 0$, it follows that r is a retraction.

The rule

$$\begin{array}{llll} H'\colon & I \times \mathbf{Homeo}_+(S^1) \times \mathbb{R} & \to & S^1 \\ & (t, f, \phi) & \mapsto & p((1-t)(\tilde{f}(\phi) - r(f)) + t\phi + r(f)) \end{array}$$

defines a continuous S^1-equivariant map which has period 1 in ϕ. It induces a continuous function

$$H\colon I \times \mathbf{Homeo}_+(S^1) \times S^1 \to S^1,$$

the adjoint of which is a map

$$\hat{H}\colon I \times \mathbf{Homeo}_+(S^1) \to \mathbf{map}(S^1, S^1),$$

which is continuous by Proposition 3.4. It remains to show that the values of $\hat{H}$ belong to $\mathbf{Homeo}_+(S^1)$.

Since every continuous map $f\colon S^1 \to S^1$ of degree 1 is surjective, see [HY61], Theorem 6-31, and closed, it suffices to show that the values of $\hat{H}$ are injective maps of degree 1. Since the degree of a map

$$\deg\colon \mathbf{map}(S^1, S^1) \to \mathbb{Z}$$

is a homotopy invariant, see [HY61], Theorem 6-29, it follows that every $\hat{H}(t, f)$ has degree 1.

We claim that $\hat{H}(t, f)$ is injective. Suppose that $0 < \phi - \phi' < 1$. Since f is a homeomorphism of degree 1, it follows that $0 < \tilde{f}(\phi) - \tilde{f}(\phi') < 1$. Hence

$$0 < (1-t)(\tilde{f}(\phi) - \tilde{f}(\phi')) + t(\phi - \phi') < 1,$$

which shows that H is injective in variable ϕ.

This completes the proof that S^1 is a strong S^1-equivariant deformation retract of $\mathbf{Homeo}_+(S^1)$. □

In this thesis the equivariant simplex $X = \Delta_n(G; K_0, \ldots, K_n)$ has been considered as a left G-space as well as a right N_0^n-space. Since the right N_0^n-action commutes with the left G-action and preserves G-orbits, it gives rise to an antihomomorphism $\psi\colon N_0^n \to \mathbf{Homeo}^G_{\Delta_n}(X)$. The main point demonstrated in this thesis has been that for G a Lie group, the group $\psi(N_0^n)$ of right translations of X is a strong equivariant deformation retract of $\mathbf{Homeo}^G_{\Delta_n}(X)$, see Theorem 8.5.

In formulating the preliminary result, Theorem 8.4, the author's primary intention has been to emphasise the relevant features of Lie groups. However, as Example 9.9 above shows, Theorem 8.4 provides a non-trivial generalisation of Theorem 8.5. Although we are not yet aware of any practical applications of this greater generality, restricting the analysis to Lie groups would not have made our arguments significantly shorter.

References

[Are46] Richard Arens, *Topologies for homeomorphism groups*, Amer. J. Math. **68** (1946), 593–610.

[Bou66] Nicolas Bourbaki, *Elements of mathematics. General topology. Part 1*, Hermann, Paris, 1966.

[Bou72] N. Bourbaki, *Éléments de mathématique. Fasc. XXXVII. Groupes et algèbres de Lie. Chapitre II: Algèbres de Lie libres. Chapitre III: Groupes de Lie*, Hermann, Paris, 1972.

[Bre72] Glen E. Bredon, *Introduction to compact transformation groups*, Pure and Applied Mathematics, vol. 46, Academic Press, New York, 1972.

[Che46] Claude Chevalley, *Theory of Lie groups. I*, Princeton University Press, Princeton, N.J., 1946.

[Dow51] C. H. Dowker, *On countably paracompact spaces*, Canadian J. Math. **3** (1951), 219–224.

[Dug66] James Dugundji, *Topology*, Allyn and Bacon, Boston, Mass., 1966.

[Hel78] Sigurdur Helgason, *Differential geometry, Lie groups, and symmetric spaces*, Pure and Applied Mathematics, vol. 80, Academic Press, New York, 1978.

[HR63] Edwin Hewitt and Kenneth A. Ross, *Abstract harmonic analysis. Vol. I: Structure of topological groups. Integration theory. Group representations*, Die Grundlehren der mathematischen Wissenschaften, Bd. 115, Springer, Berlin, 1963.

[HY61] John G. Hocking and Gail S. Young, *Topology*, Addison-Wesley, Reading, MA, 1961.

[Ill72a] Sören Illman, *Equivariant algebraic topology*, Ph.D. thesis, Princeton University, Princeton, N.J., 1972.

[Ill72b] Sören Illman, *Equivariant singular homology and cohomology for actions of compact Lie groups*, Proceedings of the Second Conference on Compact Transformation Groups (Univ. Massachusetts, Amherst, Mass., 1971), Part I, Lecture Notes in Math., vol. 298, Springer, Berlin, 1972, pp. 403–415.

[Ill78] Sören Illman, *Smooth equivariant triangulations of G-manifolds for G a finite group*, Math. Ann. **233** (1978), 199–220.

[Ill83] Sören Illman, *The equivariant triangulation theorem for actions of compact Lie groups*, Math. Ann. **262** (1983), 487–501.

[Ill00] Sören Illman, *Existence and uniqueness of equivariant triangulations of smooth proper G-manifolds with some applications to equivariant Whitehead torsion.*, J. reine angew. Math. **524** (2000), 129–183.

[Ill02] Sören Illman, *Hilbert's fifth problem and proper actions of Lie groups*, Current trends in transformation groups, *K*-Monogr. Math., vol. 7, Kluwer Acad. Publ., Dordrecht, 2002, pp. 1–23.

[Jam84] I. M. James, *General topology and homotopy theory*, Springer, New York, 1984.

[Kel55] John L. Kelley, *General topology*, D. Van Nostrand, Princeton, N.J., 1955.

[MS87] Takao Matumoto and Masahiro Shiota, *Unique triangulation of the orbit space of a differentiable transformation group and its applications*, Homotopy theory and related topics (Kyoto, 1984), Adv. Stud. Pure Math., vol. 9, North-Holland, Amsterdam, 1987, pp. 41–55.

[Spa66] Edwin H. Spanier, *Algebraic topology*, McGraw-Hill, New York, 1966.

[Ste51] N. Steenrod, *The topology of fibre bundles*, Princeton University Press, Princeton, N.J., 1951.

Department of Mathematics and Statistics
P.O. Box 68, FI-00014 University of Helsinki, Finland
E-mail: roman.goebel@helsinki.fi